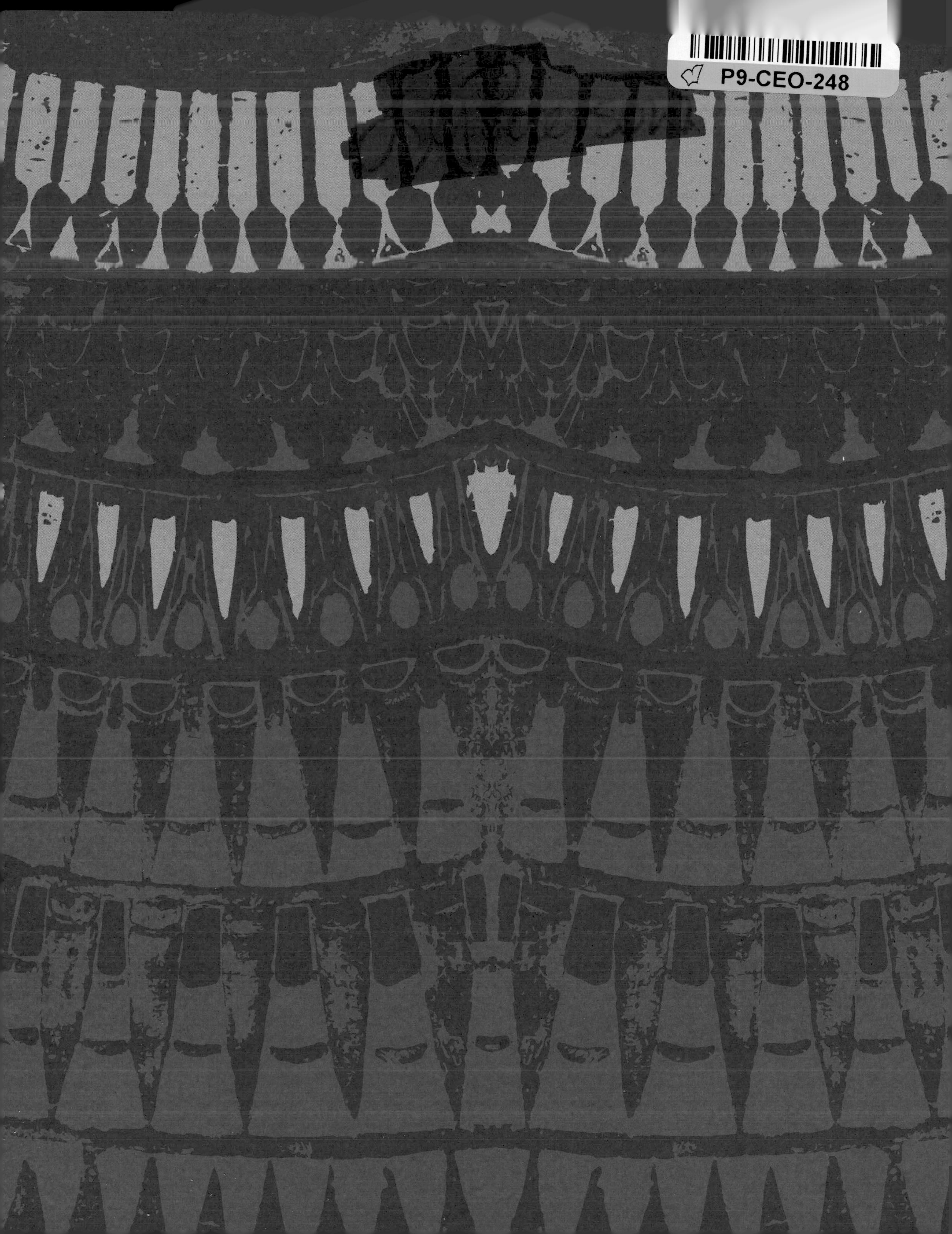
P9-CEO-248

ANCIENT
EGYPT

ANCIENT EGYPT

CIVILIZATIONS OF THE NILE VALLEY FROM PHARAOHS TO FARMERS

Bath • New York • Cologne • Melbourne • Delhi
Hong Kong • Shenzhen • Singapore • Amsterdam

This edition published by Parragon Books Ltd in 2014 and distributed by

Parragon Inc.
440 Park Avenue South, 13th Floor
New York, NY 10016
www.parragon.com

Cover image: The Art Archive / Egyptian Museum Cairo / Araldo De Luca.
Wooden throne, gilded and inlaid, showing detail of King Tutankhamun, 1332-1322 BC, detail of back, from his tomb in the Valley of the Kings, discovered by Howard Carter in 1922, 18th Dynasty.

ISBN 978-1-4723-4902-6

Printed in China

CONTENTS

INTRODUCTION

Ancient Egypt, one of the oldest civilizations on the planet, still intrigues and amazes researchers to this day. As archaeological excavations, involving the use of the most modern forms of technology available, progress through the depths of the desert, new evidence of grandeur emerges, of which much still remains to be discovered. The existing information constantly transforms into new mysteries and new grounds for research and scientific, philosophical, and artistic speculation.

There are few cultures like Ancient Egypt that have been condensed into such a compact and closed-off world, despite it having been tied to other settlements in Asia, Africa, and Europe throughout its history. Egypt was often a conqueror of other lands and, in turn, was conquered. However, none of these changes resulted in the culture abandoning its deep sense of idiosyncrasy, forged around 3000 BCE in the Nile River Valley. Its pyramids, converted into the emblematic symbol of the land of the Pharaohs, represent the monumental expression of a particular view of the world.

In Ancient Egypt, a hierarchically structured social model, based on castes and governed rigidly through these layers, reigned supreme. It was headed by the figure of the Pharaoh, located at the highest seat of power, and descended through to a wide range of peasants and slaves. The bond that tied this social "pyramid" together was a religious belief that, after integrating several regional pantheons, was built around a single figure—the Pharaoh himself—who enjoyed a dual nature: both human and divine. The immense monumental nature of Egyptian architecture and sculpture embodied this

Zoomorphic funerary statuette, deposited by the Ancient Egyptians in the tombs of their dead to help their spirits in the transition to the afterlife.

dedication to grandeur and eternity, which encompassed both life on earth and the afterlife.

However, this omnipotence did not suffocate or paralyze the Ancient Egyptians. On the contrary, they revolutionized the technology of their time, developed science to astounding heights, and expressed themselves artistically through a series of unmistakable aesthetic structures. Along with the way that pyramids and obelisks became universal symbols known to cultures around the world, the entire planet is also indebted to the Ancient Egyptians—for their agrarian and construction techniques; their significant mathematical knowledge; their development of the field of medicine and their expertise in surgery; the beauty of their statues, bas-relief sculptures, and wall paintings as well as their excellent classical literature and the depth of their philosophical contemplations. The footprint of Ancient Egypt can be perceived in many of the main religious and speculative beliefs supported around the world today.

And it all began, naturally, on the shores of a river—the sacred Nile. The frequency of its swells, the rate of its floods, and the fertility of its silt fed, cradled, and helped the civilization lay its roots like no other. Never has a civilization withstood the onslaught of time quite like the Egyptians.

Sphinx, constructed around 2050–1786 BCE, a significant, and yet mysterious, symbol.

CHRONOLOGY

The history of Ancient Egypt dates back to time immemorial. Despite the passing of the centuries and its various transformations, the Egyptian Kingdom retained its own profile. Even when occupied by outside powers, such as Persia, Greece, and Rome, it assimilated these foreign cultures, stamping its own footprint on all of them. Historians use the various successive dynasties that governed the kingdom to distinguish the different periods and describe its evolution. However, dating the dynasties remains a point of contention to this day. This book has used the *Heilbrunn Timeline of Art History*, from the Metropolitan Museum of Art in New York, as a point of reference.

ca. 4500 BCE

PREDYNASTIC

Zekhen

3100 BCE

EARLY DYNASTIC

3100–2770 BCE

1ST DYNASTY

Narmer or Menes
Djer
Djet
Den
Anedjib
Semerkhet
Qa'a

2770–2649 BCE

2ND DYNASTY

Hotepsekhemwy
Raneb or Nebra
Nynetjer
Peribsen
Khasekhem /Khasekhemwy

2649 BCE

OLD KINGDOM

2649–2575 BCE

3RD DYNASTY

Sanakhte (2649–2630 BCE)
Djoser or Zoser (2630–2611 BCE)
Sekhemkhet (2611–2603 BCE)
Khaba (2603–2599 BCE)
Huni (2599–2575 BCE)

Chariots of war. Brought to Ancient Egypt by the Hyksos, along with bronze-tipped arrows and the compound Asian bow.

2575–2465 BCE

4TH DYNASTY

Sneferu (2575–2551 BCE)
Khufu or Cheops (2551–2528 BCE)
Djedefra (2528–2520 BCE)
Khafra or Chephren (2520–2494 BCE)
Nebka II (2494–2490 BCE)
Menkaure or Mykerinos (2490–2472 BCE)
Shepseskaf (2472–2467 BCE)
Thampthis (2467–2465 BCE)

2465–2323 BCE

5TH DYNASTY

Userkaf (2465–2458 BCE)
Sahure (2458–2446 BCE)
Neferirkare Kakai (2446–2438 BCE)
Shepseskare Isi (2438–2431 BCE)
Neferefre (2431–2420 BCE)
Nyuserre Ini (2420–2389 BCE)
Menkauhor Kaiu (2389–2381 BCE)
Djedkare Isesi (2381–2353 BCE)
Unas (2353–2323 BCE)

2323–2150 BCE

6TH DYNASTY

Teti (2323–2291 BCE)
Userkare (2291–2289 BCE)
Pepi I (2289–2255 BCE)
Merenre I (2255–2246 BCE)
Pepi II (2246–2152 BCE)
Merenre II (2152–2152 BCE)
Neitiqerty Siptah (2152–2150 BCE)

2150 BCE

FIRST INTERMEDIATE

2150–2124 BCE

7TH TO 10TH DYNASTIES

Neferkara, Khety (among various transitional, very unstable Pharaohs).

Pharaoh Pepi II. Alabaster statue, in which Pepi II (6th Dynasty) appears as a child on the lap of his mother, Queen Ankhnes-meryre.

From the first to the eighth Dynasty

The first pyramid rises on the horizon

After uniting the kingdoms of Upper and Lower Egypt, King Menes (or Narmer) established the capital in Memphis and the Ancient Egyptian Kingdom was born. The calendar was invented, hieroglyphs were developed and the land of the Nile took giant steps forward, from consolidating its political power to writing the first treatise on surgery. These scientific advances were also accompanied by extraordinary economic development, as a result of improved agricultural techniques with advanced irrigation and drainage. Between 2630 and 2528 BCE, the architect Imhotep designed the first pyramid, the stepped Saqqara pyramid, and Khufu, the Great Pyramid in Giza. The city of Heliopolis grew in stature, and its god Re was inaugurated into the pantheon. Despite local resistance, the Pharaoh imposed a centralized, vertical, social model.

2124–2030 BCE

11TH DYNASTY (FIRST HALF)

Mentuhotep I (2124–2120 BCE)
Intef I (2120–2108 BCE)
Intef II (2108–2059 BCE)
Intef III (2059–2051 BCE)
Mentuhotep II (2051–2030 BCE)

2030 BCE

MIDDLE KINGDOM

2030–1981 BCE

11TH DYNASTY (SECOND HALF)

Mentuhotep II cont. (2030–2000 BCE)
Mentuhotep III (2000–1988 BCE)
Qakare Intef (ca. 1985 BCE)
Sekhentibre (ca. 1985 BCE)
Menekhkare (ca. 1985 BCE)
Mentuhotep IV (1988–1981 BCE)

1981–1802 BCE

12TH DYNASTY

Amenemhat I (1981–1952 BCE)
Senusret I (1961–1917 BCE)
Amenemhat II (1919–1885 BCE)
Senusret II (1887–1878 BCE)
Senusret III (1878–1840 BCE)
Amenemhat III (1859–1813 BCE)
Amenemhat IV (1814–1805 BCE)

1802–1640 BCE

13TH DYNASTY

1640 BCE

SECOND INTERMEDIATE

1640–1635 BCE

14TH TO 16TH DYNASTIES

1635–1550 BCE

17TH DYNASTY

Seqenenre Tao I (ca. 1560 BCE)
Seqenenre Tao II (ca. 1560 BCE)
Kamose (ca. 1552–1550 BCE)

1550 BCE

NEW KINGDOM

1550–1295 BCE

18TH DYNASTY

Ahmose I (1550–1525 BCE)
Amenhotep I (1525–1504 BCE)
Thutmose I (1504–1492 BCE)
Thutmose II (1492–1479 BCE)
Thutmose III (1479–1425 BCE)
Hatshepsut (as regent) (1479–1473 BCE)
Hatshepsut (1473–1458 BCE)
Amenhotep II (1427–1400 BCE)
Thutmose IV (1400–1390 BCE)
Amenhotep III (1390–1352 BCE)
Amenhotep IV / Akhenaten (1352–1336 BCE)
Nefertiti (1338–1336 BCE)
Smenkhkare (1336 BCE)
Tutankhamun (1336–1327 BCE)
Sahure (1327–1323 BCE)
Horemheb (1323–1295 BCE)

1295–1186 BCE

19TH DYNASTY

Ramesses I (1295–1294 BCE)
Seti I (1294–1279 BCE)

Gigantic sculpture. Statue of Khafre, a clear example of the monumental style of Egyptian sculptures. It dates back to the twenty-sixth century BCE. The wings of a falcon extend from its back, a zoomorphic representation of the god Horus.

From the ninth to the twelfth Dynasty

Conflict between Upper and Lower Egypt

In 2134 BCE, the ninth Dynasty came to power. Its excessively centralized administration angered the representatives of the nomes—feudal lords with significant local interests. Tensions between Upper and Lower Egypt intensified; Nubian tribes took advantage of this, and were ready to undermine the power of the kingdom. The government of the tenth Dynasty failed to rectify the situation, so chaos and hunger reigned supreme, with looting of the temples commonplace. In 2030 BCE, Mentuhotep II, representative of the eleventh Dynasty, reunified Upper and Lower Egypt and, promoting the cult of Amun, transferred the capital to Thebes. The economic recovery fostered the construction of monumental buildings. The next dynasty promoted the cult of Osiris, as god of the dead.

Ramesses II (1279–1213 BCE)
Merenptah (1213–1203 BCE)
Amenmesse (1203–1200 BCE)
Seti II (1200–1194 BCE)
Siptah (1194–1188 BCE)
Tawosret (1188–1186 BCE)

1186–1070 BCE

20TH DYNASTY

Setnakht (1186–1184 BCE)
Ramesses III (1184–1153 BCE)
Ramesses IV (1153–1147 BCE)
Ramesses V (1147–1143 BCE)
Ramesses VI (1143–1136 BCE)
Ramesses VII (1136–1129 BCE)
Ramesses VIII (1129–1126 BCE)
Ramesses IX (1126–1108 BCE)
Ramesses X (1108–1099 BCE)
Ramesses XI (1099–1070 BCE)

1070 BCE

THIRD INTERMEDIATE

1070–945 BCE

21ST DYNASTY

Smendes I (1070–1044 BCE)
Amenemnisu (1044–1040 BCE)
Psusennes I (1040–992 BCE)
Amenemope (993–984 BCE)
Osorkon (984–978 BCE)
Siamun (978–959 BCE)
Psusennes II (959–945 BCE)

945–712 BCE

22ND DYNASTY (LIBYAN)

Shoshenq I (945–924 BCE)
Osorkon I (924–889 BCE)
Shoshenq II (ca. 890 BCE)
Takelot I (889–874 BCE)
Osorkon II (874–850 BCE)
Harsiese (ca. 865 BCE)
Takelot II (850–825 BCE)
Shoshenq III (825–773 BCE)
Pami (773–767 BCE)
Osorkon IV (730–712 BCE)

818–713 BCE

23RD DYNASTY

Pedubast I (818–793 BCE)
Iuput I (ca. 800 BCE)
Shoshenq IV (793–787 BCE)
Osorkon III (787–759 BCE)
Takelot III (764–757 BCE)
Rudamun (757–754 BCE)
Iuput II (754–712 BCE)
Peftjauwybast (740–725 BCE)
Namlot (ca. 740 BCE)
Thutemhat (ca. 720 BCE)

724–712 BCE

24TH DYNASTY

Tefnakht (724–717 BCE)
Bakenrenef (717–712 BCE)

712 BCE

LATE PERIOD

712–664 BCE

25TH DYNASTY (NUBIAN)

Piye (743–712 BCE)
Shabaka (712–698 BCE)
Shebitku (698–690 BCE)
Taharqa (690–664 BCE)
Tantamani (664–653 BCE)

688–525 BCE

26TH DYNASTY (SAITE)

Nikauba (688–672 BCE)
Necho I (672–664 BCE)
Psamtik I (664–610 BCE)
Necho II (610–595 BCE)
Psamtik II (595–589 BCE)
Apries (589–570 BCE)
Ahmose II (570–526 BCE)
Psamtik III (526–525 BCE)

525–404 BCE

27TH DYNASTY (1ST ACHAEMENID PERIOD)

Cambyses (525–522 BCE)
Darius I (521–486 BCE)
Xerxes I (486–466 BCE)
Artaxerxes I (465–424 BCE)
Darius II (424–404 BCE)

522–399 BCE

28TH DYNASTY

Pedubast III (522–520 BCE)
Psamtik IV (ca. 470 BCE)
Inaros (ca. 460 BCE)
Amyrtaeus I (ca. 460 BCE)
Thannyros (ca. 445 BCE)
Pausiris (ca. 445 BCE)
Psamtik V (ca. 445 BCE)
Psamtik VI (ca. 400 BCE)
Amyrtaeus II (404–399 BCE)

Princess. Bust of an Egyptian princess from the eighteenth Dynasty, dating to around the fourteenth century BCE.

The serpent king. A stone stele that dates back to the third millennium BCE as a tribute to the "serpent king," one of the successors of Menes. It measures 4 ft 8 in/1.43 m in height and was found at the necropolis of Abydos.

From the thirteenth Dynasty to decadence

After a wave of invasions, the Roman conquest

Over the course of the eighteenth century BCE, during the thirteenth Dynasty, Hyksos invaders made progress into the kingdom and ascended to the seat of power. Nonetheless, they were characterized by their significant contributions to Egyptian culture, from new sewing techniques to chariot models, from the lyre to the lute. Thanks to this recovery, the New Kingdom was created in 1550 BCE. Ahmose expelled the Hyksos, and Akhenaten built a new capital at Tell el-Amarna. The dynasties that followed, until the thirty-first ended, were unable to preserve pharaonic power. The Nubians, Greeks, and Persians successively conquered the territory, and in 332 BCE Alexander the Great conquered Egypt and founded Alexandria. His successors began a period of great political weakness, but of spectacular cultural wealth. Then, in 30 BCE, Egypt fell into the hands of Rome and never recovered.

399–380 BCE

29TH DYNASTY

Nepherites I (399–393 BCE)
Psammuthes (393 BCE)
Achoris (393–380 BCE)
Nepherites II (380 BCE)

380–343 BCE

30TH DYNASTY

Nectanebo I (380–362 BCE)
Teos (365–360 BCE)
Nectanebo II (360–343 BCE)

343–332 BCE

31ST DYNASTY (2ND ACHAEMENID PERIOD)

Artaxerxes III (343–338 BCE)
Arses (338–336 BCE)
Darius III (335–332 BCE)

332 BCE

HELLENISTIC PERIOD

332–304 BCE

MACEDONIAN DYNASTY

Alexander the Great (332–323 BCE)
Philip III Arrhidaeus (323–316 BCE)
Alexander IV (316–304 BCE)

304–30 BCE

PTOLEMAIC DYNASTY

Ptolemy I Soter (304–284 BCE)
Ptolemy II Philadelphos (285–246 BCE)
Arsinoe II (278–270 BCE)
Ptolemy III Euergetes (246–221 BCE)
Berenice II (246–221 BCE)
Ptolemy IV Philopator (221–205 BCE)
Ptolemy V Epiphanes (205–180 BCE)
Hugronaphor (205–199 BCE)
Ankhmakis (199–186 BCE)
Cleopatra I (194–176 BCE)
Ptolemy VI Philopator (180–145 BCE)
Cleopatra II (175–115 BCE)
Ptolemy VIII Euergetes II (170–116 BCE)
Harsiese (ca. 130 BCE)
Ptolemy VII Neos Philopator (163–145 BCE)
Cleopatra III (142–101 BCE)
Ptolemy IX Soter II (116–80 BCE)
Ptolemy X Alexander I (107–88 BCE)
Ptolemy XI Alexander II (88–80 BCE)
Ptolemy XII Neos Dionysos (80 51 BCE)
Cleopatra IV Berenice III (ca. 79 BCE)
Berenice IV (58–55 BCE)
Ptolemy XIII (51–47 BCE)
Cleopatra VII (51–30 BCE)
Ptolemy XIV (47–44 BCE)
Ptolemy XV (44–30 BCE)

30 BCE–364 CE

ROMAN PERIOD

364–476 CE

BYZANTINE PERIOD

The Triad of Menkaure. One of the many triads of Menkaure, who appears with the goddess Hathor and a local god. It dates back to the twenty-fifth century BCE.

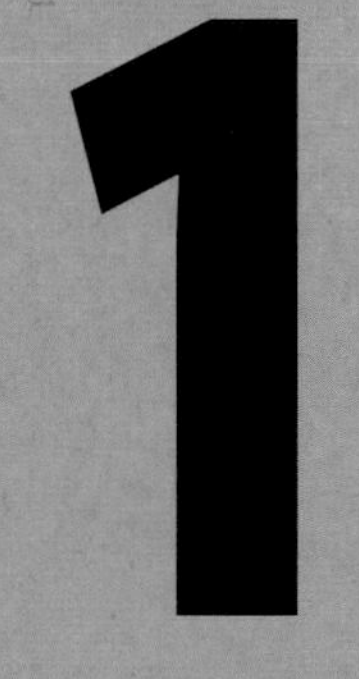

CHAPTER

FASCINATED BY EGYPT

THE HISTORY OF EGYPTOLOGY

Ancient Egypt has always provoked great fascination among all later civilizations. However, it was not until the end of the eighteenth century that it was studied scientifically, giving rise to a new discipline—Egyptology.

Ancient Egypt is one of the most fascinating civilizations of all time. Greeks, Romans, Arabs, adventurers, scientists, and modern thinkers have been attracted by this civilization that, to them, appears almost perfect and yet mysterious: perfect, as its social organization remained unshakable, with barely any social conflict, for almost 3,000 years; mysterious, based on its unintelligible hieroglyphic writing style, the omnipresent figure of the Pharaoh, the meaning and process of building the immense pyramids and temples, and its magnificent tombs and mummies.

The origin of Egyptology

The passion for Egypt dates almost as far back as the civilization itself. The subject has always inspired a desire to understand and discover through study and research. An interest in Egypt's history and society was demonstrated by ancient historians such as Herodotus, Dionysius I of Syracuse, and Manetho. The chronological division of the Pharaohs among the different dynasties owes much to Manetho and his great work, *Aegyptiaca*, written in the third century BCE. Also, since ancient times there has been an interest in owning and imitating Egyptian art and architecture. In the Roman era, villas such as Villa Adriana (Hadrian's Villa), at Tivoli, imitated Egyptian architecture and contained objects recovered from the land of the Nile. Pyramids and obelisks were also reproduced in Roman temples and squares, such as the Pyramid of Cestius from the Augustan period.

While little progress was made during the Middle Ages on the study of Egypt, from the seventeenth century onward the fascination was reborn with the first scientific attempts to investigate the keys to this great

TUTANKHAMUN AND HIS WIFE Detail of Tutankhamun's throne, found in his tomb, whose discovery in 1922 represents a milestone in Egyptology.

ART FROM AMARNA
Relief from the Amarna period (1353–1336 BCE). The excavations at Tell el-Amarna led to the discovery of an artistic renaissance during the reign of Akhenaten.

civilization. This movement gained momentum during the eighteenth century, and reached its peak at the end of that century. The land of the pyramids captured the minds of the Western world once again. However, it still proved impossible to decode its secrets.

The Napoleonic Expedition

The 1798 military campaign in North Africa undertaken by Napoleon Bonaparte, who embarked on his journey with a large team of academics, marked the turning point in the approach to Ancient Egypt. The scientific arm of this expedition, formed by members of the Commission des Sciences et des Arts, was dedicated to studying all aspects of Modern and Ancient Egypt. Renowned experts from different disciplines, such as Vivant Denon, Claude-Louis Berthollet, Gaspard Monge, and Joseph Fourier, worked in the field, collecting data and illustrating everything they found. Their headquarters was the Institut d'Égypte in Cairo, created by Napoleon as a mirror image of the French Institute. The fruits of this monumental task could be found in *Description de l'Égypte*, published in Paris between 1809 and 1828, drafted by the Commission des Sciences et des Arts under the guidance of Denon. The great expedition promoted by Napoleon was the first of many that would take place thereafter.

TUTANKHAMUN PECTORAL
One of the jewels found in Tutankhamun's tomb, made from precious and semiprecious stones. Currently, the Pharaoh's treasure can be found in the Cairo Museum.

Deciphering the hieroglyphs

The discovery of monuments facilitated a greater understanding of Egypt. However, there were mysteries left to be discovered, and one of the most intriguing was the hieroglyphic writing system. Among the various unsuccessful attempts to decipher it, the work of Jesuit scholar Athanasius Kircher is particularly renowned. During the mid-seventeenth century, he published four volumes of translations that would prove completely inaccurate.

It was a chance discovery during the Napoleonic Expedition that held the key to moving the study of hieroglyphic writing forward: the Rosetta Stone

The discovery of the Rosetta Stone in the Nile delta happened by chance during the Napoleonic Expedition—it held the key to moving forward in the study of the writing style and, as a result, Egyptian civilization. The importance of this stone, which contained the same inscription carved in hieroglyphs, demotic (the cursive form of hieroglyphs), and Greek, was clear to French researchers, who concentrated on studying it and sent copies to academics throughout Europe. The discovery was quickly transformed into a dispute between the British and the French, rivals on both the battlefield and in science. The British, who defeated Napoleon's troops in 1801 in Alexandria and who controlled the Mediterranean, took control of the stone and sent it to England. It was clear that the granite stone held the key to unlocking the mystery of the hieroglyphs, but it would take a number of years to be fully resolved. Englishman Thomas Young made progress and discovered that it was not an alphabetic writing system, but it was French philologist Jean-François Champollion who, beating his English rival to the discovery, managed to decipher it in 1822, using a copy that had reached him in 1808. The foundations of Ancient Egypt's writing system were therefore established, having been used for the last time during the fifth century CE. This discovery finally made it possible to fully understand many of the wonders found that had, until that time, remained a mystery.

Thus, Napoleon's expedition was doubly important. On the one hand, his campaign for the first time added a scientific viewpoint to expeditions; on the other, with the discovery of the Rosetta Stone, it was possible to make huge steps forward in understanding this ancient civilization. Researchers were finally able to read the hieroglyphs that feature in temples and pyramids, and on stones and papyrus. Thus, this period is considered the starting point of modern Egyptology, the scientific discipline dedicated to the study of Ancient Egypt.

STATUETTE OF TUY
Gilded statuette of grenadilla and shea wood belonging to the eighteenth Dynasty during the New Kingdom. Today, it is in the collection of the Louvre Museum in Paris.

PRIESTESS HENT-MAUI'S PAPYRUS
Representation of a prayer to the god Re, with hieroglyphic writing on a papyrus from the twenty-first Dynasty, around 1000 BCE, which is in the British Museum in London.

SERAPEUM STELE
Funerary stele from Serapeum of Saqqara (necropolis of the Apis bulls). The sarcophagi of the sacred animals were found empty, but they featured stelae with information about each bull.

The great expeditions of the nineteenth century

In the nineteenth century, Egyptology made a massive leap and the number of excavations and discoveries increased. Between 1815 and 1819, explorer Giovanni Battista Belzoni made important findings, including the entrances to the temple of Abu Simbel and the pyramid of Khafre (Chephren, in Greek), which made access to them possible for the first time. He discovered six tombs in the Valley of the Kings, among which was the tomb belonging to Seti I, one of the most impressive of the entire burial site. In 1828, Jean-François Champollion accompanied his Italian colleague Ippolito Rosellini on a French-Tuscan expedition that toured Egypt for a year, compiling documentation and providing material for the Egyptian collections at the Louvre Museum and the Egyptian Museum in Florence. Champollion died shortly after returning from the expedition, while Rosellini published a detailed, impressive work describing the results of his research.

Both men were succeeded by Richard Lepsius, who led the Prussian expedition to Egypt from 1842 to 1845, and who contributed to the creation of the Egyptian Museum in Berlin, of which he was the director. In 1866, he found the Decree of Canopus on the outskirts of Tanis; this text, written in Greek, demotic, and hieroglyphs during the third century, confirmed Champollion's translations.

Egyptology, as a scientific discipline, gained strength in the following years. At the end of the nineteenth century and beginning of the twentieth century, expeditions from France, Italy, Germany, and the United Kingdom multiplied, making significant discoveries, each of which caused great repercussions both in public opinion and among the scientific community. The works of Émile Brugsch are particularly renowned; in 1881, he discovered the mummies of the New Kingdom Pharaohs at Deir el-Bahri, including those of Thutmose III and Ramesses II.

The qualitative leap in Egyptology was thanks to Britain's William Matthew Flinders Petrie, who was the first to use a genuinely scientific research method in archaeological excavations, which he carried out over a 40-year period, classifying all his findings and systematically publishing his investigations.

Archaeological looting

During the first half of the nineteenth century, and thanks to numerous expeditions, great collections in the main European museums were created and Egyptian museums were founded in various cities across Europe. These institutions embarked on a frenetic race to acquire huge quantities of objects brought from the Nile region by expeditions and foreign diplomats residing in Egypt—in reality, this represented large-scale looting of archaeological heritage. It was François Auguste Mariette, in particular, who discovered the Serapeum at Saqqara (the necropolis of the Apis bulls of Memphis), and who reversed this trend, choosing to leave the archaeological heritage in Egypt itself. Mariette founded the Boulaq Museum, the predecessor of the present-day Museum of Egyptian Antiquities, in order to preserve the findings of the numerous archaeological expeditions embarked upon by different countries.

In addition, the discoveries and scientific advances fed Egyptology fever in Europe and the Western world, which continued throughout the nineteenth century. Decorative art, in addition to architecture, was widely inspired by pharaonic art. Hotels, banks, libraries, and other public and commercial buildings in New York and European capitals were filled with ornaments that imitated the style of Ancient Egypt, in addition to squares, parks, and streets in which monuments emulating obelisks and pyramids were erected, while mausoleums in the style of ancient temples became widespread in cemeteries.

During the nineteenth century, great collections in the main European museums were created and Egyptian museums were founded in various European cities

QUEEN NEFERTARI'S TOMB
Wall painting from the funerary chamber of Queen Nefertari (nineteenth Dynasty), wife of Ramesses II, located in the Valley of the Queens.

MASK OF SHOSHENQ II
Made from gold, it was one of the pieces found in the necropolis of Tanis. It belonged to Shoshenq II, of the twenty-second Dynasty.

TANIS TREASURES
Armlet and golden vase, two of the valuable finds in the tomb of Psusennes, one of the tombs at the necropolis of Tanis.

RAMESSES II
Statuette of Ramesses II, wearing a blue crown. At his feet, and much smaller in size, is a representation of his wife, Nefertari.

Discoveries at the beginning of the twentieth century

During the last century the fever continued and expeditions ensued. First, Italian Ernesto Schiaparelli, director of the Egyptian Museum in Florence and later of its counterpart in Turin, discovered around 80 tombs in the Valley of the Queens, including that of Nefertari, wife of Ramesses II, between 1903 and 1905. Ludwig Borchardt led important excavations at the Tell el-Amarna site, a city created by Akhenaten as part of his religious reforms, where, in 1912, the famous bust of Queen Nefertiti was discovered. In 1939, Pierre Montet discovered the necropolis of Tanis intact, belonging to kings, family members and other dignitaries from the sixteenth and seventeenth Dynasties.

However, the most famous of the discoveries, without a doubt, was that of the young Pharaoh Tutankhamun, in the Valley of the Kings in 1922. Englishman Howard Carter, hungry for a notable discovery, excavated a sectioned-off area of the Valley of the Kings for eight years. His persistence was rewarded. The story of the discovery of the tomb, the intact burial paraphernalia, and the mummy of the Pharaoh itself, stoked up worldwide admiration and caused a new wave of Egyptomania, which wasn't lacking in esoteric and mysterious explanations—as reflected in the jewels, furniture, and decorative objects from the period.

A long way to go

Excavation work and research have continued to this day, and they still produce surprising discoveries: the tomb of the children of Ramesses II in the Valley of the Kings (1987); the necropolis of Abydos, with ditches to house boats from the Predynastic period (1991); the necropolis featuring hundreds of mummies from the Hellenistic period in the Bahariya Oasis (1996); a new passage in the Great Pyramid with two doors, the meaning of which remains unknown (2002); the city and the necropolis of the men who

Over the past couple of decades, archaeological excavations have benefited from new technologies, such as radar, satellite imagery, robots, and genetic analysis

built the Pyramids of Giza (1990) or the 83 mummified animals discovered in Abydos (2012), to name just a few of the most famous recent discoveries.

Over the past couple of decades, archaeological excavations have benefited from technological advances. Thanks to radar, scanners, robots, satellite imagery, genetic analysis, and other techniques, it has been possible to take large steps forward in the understanding of this ancient civilization, in addition to deciphering some of the mysteries that had remained unsolved. The work of Egyptian Egyptologist Zahi Hawass, head of the Supreme Council of Antiquities in Egypt, is particularly renowned; he led research into the death of Tutankhamun using scans and comparing his DNA with those of others, to establish his parentage. Using the results of his work, he was able to disprove many existing theories about the death of the young Pharaoh and identify mummies that belonged to his family.

The efforts of modern Egyptology do not solely focus on finding and identifying monuments to document Egyptian history. They also seek to understand the structure of Egyptian society. Thanks to the joint efforts of a large number of investigators, an increasingly better knowledge of the daily lives of its inhabitants, their beliefs, organization, economy, and other aspects has been acquired.

Even so, many mysteries are yet to be uncovered. To cite just one example: 4,500 years after the construction of the Pyramids of Giza, their meaning is still not fully clear, nor are the different phases of their construction understood, nor how many chambers or passages remain hidden behind their walls... It is also estimated that only a third of the archaeological remains hidden in the desert and underneath buildings have been discovered to date. So it comes as little surprise that Ancient Egypt continues to surprise, feeding our curiosity and deserving our admiration.

CANOPIC COFFIN OF TUTANKHAMUN
One of the four miniature caskets that contained the Pharaoh's viscera. They are reproductions of the coffin that contained the mummy.

Main Sites

For 200 years, archaeological excavations in the Nile Valley have primarily focused on its capitals, funerary complexes, and royal burial sites.

Pharaohs and monuments

The most important archaeological sites in chronological order can be found in: the Protodynastic city of Abydos, home of the great temple of Osiris and the oldest royal cemetery in the world; Memphis, and the necropolises at Dahshur, Saqqara, and Giza, home to the most famous pyramids; and Thebes, home to the temples of Karnak and Luxor and the Valley of the Kings. The excavations and research undertaken here made it possible to trace the great Pharaohs who governed Ancient Egypt.

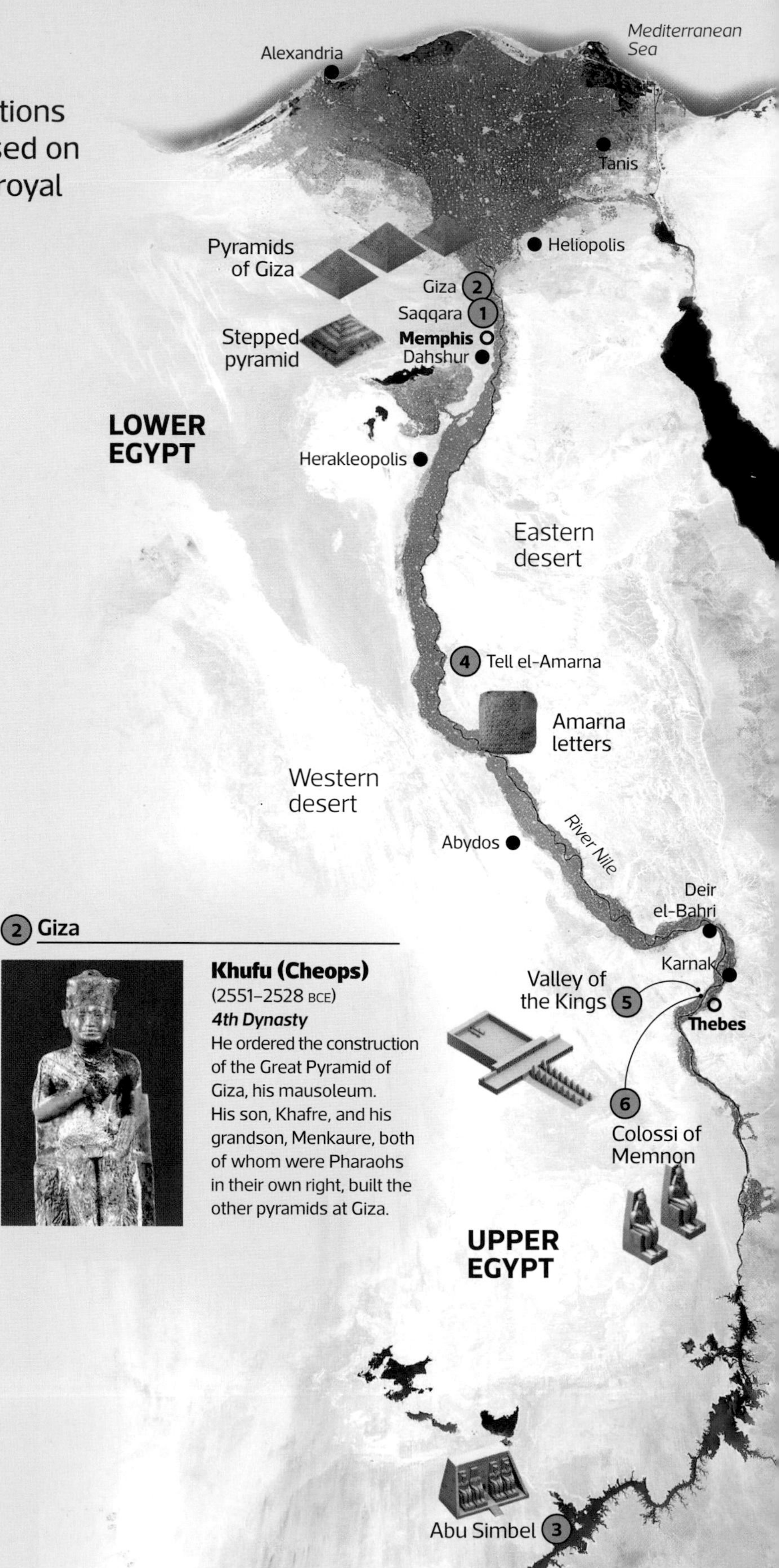

1 Saqqara

Djoser (Zoser)
(2630–2611 BCE)
3rd Dynasty
During his reign, he built the largest stone funeral complex in the world: the stepped pyramid at Saqqara.

3 Abu Simbel

Ramesses II
(1279–1213 BCE)
19th Dynasty
He multiplied the number of colossal statues, building the temples at Abu Simbel and the Ramesseum, in addition to founding the city of Pi-Ramesses, his capital.

2 Giza

Khufu (Cheops)
(2551–2528 BCE)
4th Dynasty
He ordered the construction of the Great Pyramid of Giza, his mausoleum. His son, Khafre, and his grandson, Menkaure, both of whom were Pharaohs in their own right, built the other pyramids at Giza.

4 Tell el-Amarna

Akhenaten
(1352–1336 BCE)
18th Dynasty
He promoted the monotheistic cult of Aten, installing his capital at Tell el-Amarna, a key site in the understanding of this period in history.

Nefertiti
(1338–1336 BCE)
18th Dynasty
The great wife of Akhenaten, whom he made coregent. Her bust, found at the site in Amarna, is one of the most famous.

5 Valley of the Kings

Thutmose I
(1504–1492 BCE)
18th Dynasty
Expanded the Kingdom from Upper Nubia and waged war on the banks of the Euphrates. Founder of the Royal Theban necropolis at the Valley of the Kings.

Hatshepsut
(1473–1458 BCE)
18th Dynasty
Heir to Thutmose I, the most powerful female Pharaoh in Egypt's history. She ordered the construction of the Djeser-Djeseru temple, one of the greatest architectural jewels of Ancient Egypt.

Thutmose III
(1479–1425 BCE)
18th Dynasty
Known as "the Great," he oversaw the kingdom at its greatest extension. He reigned over kingdoms in the Middle East and Eastern Mediterranean. He built the temple of Amun-Re at Karnak.

Tutankhamun
(1336–1327 BCE)
18th Dynasty
Son of Akhenaten, he restored the Cult of Amun. The most famous Pharaoh as a result of the intact discovery of his hypogeum (underground vault) in 1922 in the Valley of the Kings.

Red Sea

EGYPT

Expanded area

6 Colossi of Memnon

Amenhotep III
(1390–1352 BCE)
18th Dynasty
His reign stands out as one of the busiest construction periods in Ancient Egypt. The Colossi of Memnon are the remains of an amazing temple built next to the Nile.

Historical Sources

For centuries, the accounts of ancient historians were practically the only source that provided an insight into the history of Ancient Egypt. Occasionally, they were based on Egyptian king lists, which became accessible after hieroglyphs were deciphered.

Pioneering historians

Manetho and Herodotus are two of the main ancient historians who dedicated their lives to the study of Egypt. Manetho, an Egyptian historian and priest who lived in the third century BCE, published the work *Aegyptiaca* (History of Egypt), which detailed the dynastic divisions used in modern Egyptology. His sources include the king lists—in particular, the Turin King List. An expert on hieroglyphs, he wrote his works in Greek.

The king lists

Various king lists have been found; those from Karnak, Abydos, and Saqqara are particularly renowned. Carved into the walls of temples, they provided the chronological order of the Pharaohs, paying homage to the legitimate kings who preceded the sovereign in power at any given time.

The king list of Seti I Found on a wall in the Pharaoh's temple at Abydos. It contains a list of 76 kings, from Menes to Seti I.

HERODOTUS OF HALICARNASSUS
As part of the narratives of his travels, Herodotus (fifth century BCE) described the geography, life, and customs of the Egyptian people, in addition to providing data, some accurate and some not so, about the history of this ancient civilization.

CONSTRUCTION OF THE PYRAMIDS
According to Herodotus' reports, the pyramids were constructed using machines, similar to pulleys, that served to raise the blocks from one step to another. It is now known that this suggestion was inaccurate.

Other sources

There are two other lists that are of significant historical value. The Turin King List is a papyrus from the time of Ramesses II, written in hieratic style. It originally contained a list of kings from the divine dynasties up to the nineteenth Dynasty, ordered into columns, along with the duration of each reign. Due to its poor condition, only 90 names are legible. The second, the Palermo Stone, is one of the oldest written sources preserved in Egyptology.

Palermo Stone
Only seven fragments of a much larger block remain. It contains the names of the kings of Egypt from prehistoric times through to the fifth Dynasty. It also details the most important events of each year, including the flooding of the River Nile.

The king list of Ramesses II Located in his temple at Abydos. It contains a chronological list of 77 kings, from Menes to Ramesses II.

The Napoleonic Expedition

In May 1798, Napoleon traveled to Egypt with the aim of delaying British expansion in the Mediterranean. The campaign was a complete military failure, but resulted in successes in the fields of science and philosophy.

Relief at Dendera. Part of the Hathor temple. Illustration from the fourth volume of *Description de l'Egypte*, dedicated to antiquities.

Scientific mission

In 1798, the Commission des Sciences et des Arts was created. Around 141 of its 167 members, including renowned engineers, architects, mathematicians, astronomers, doctors, artists, botanists, and zoologists, traveled with Napoleon to study all aspects of Modern and Ancient Egypt in depth. It was the first time anyone had attempted a rigorous study of ancient Egyptian civilization. The headquarters of this investigation was the Institut d'Égypte, founded by Napoleon in the same year.

Expert work

Description de l'Egypte is a monumental work, comprising nine volumes of text and eleven volumes of illustrations. It is the result of three years of work undertaken by the academics who traveled to Egypt with Napoleon, examining, recording, and cataloguing all aspects of Modern and Ancient Egyptian civilization. More than 300 artists contributed to its creation. The illustrated volumes are divided into Antiquities, the Modern State, and Natural History. Printed in Paris between 1809 and 1828, it is considered, given its scope and rigor, the first work of modern Egyptology.

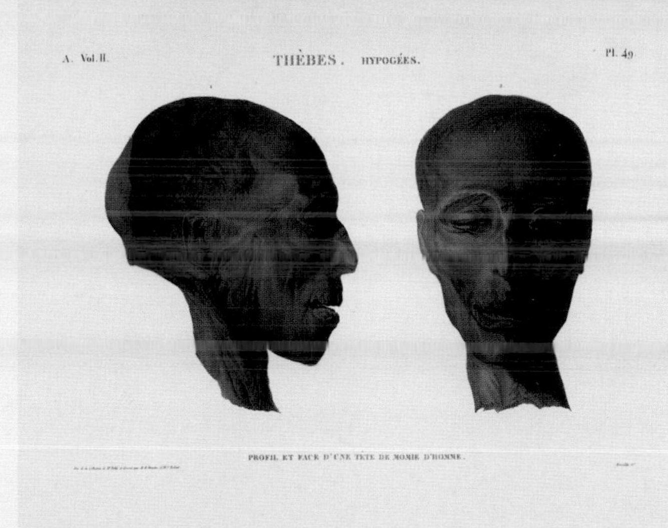

Mummy's head. This features in the second of the five volumes dedicated to antiquities. A detailed close-up of the front and profile view of a mummy.

Workers crafting bottles. In the sections dedicated to the Modern Age, numerous illustrations reflect daily life at the time.

RESEARCH IN THE FIELD
On this page, the expedition's researchers can be seen examining the Great Sphinx.

Eminent contributors

DOMINIQUE VIVANT DENON (1747–1825)
Considered the pioneer of museology and the first great Egyptologist. He led the Commission des Sciences et des Arts, in addition to discovering and illustrating Thebes, Karnak, and Aswan. His illustrations form an important part of the *Description de l'Egypte*.

CLAUDE-LOUIS BERTHOLLET (1748–1822)
A chemist and member of the Académie des Sciences, he formed part of the Physics department at the Institut d'Égypte. There, he studied the properties of natural hydrated sodium carbonate and created his theory on chemical compounds.

GASPARD MONGE (1746–1818)
A mathematician, he was the founder of the École Polytechnique. He was president of the Institut d'Égypte following its foundation. Among his contributions to the field of mathematics, the invention of descriptive geometry is worth particular mention.

JEAN BAPTISTE J. FOURIER (1768–1830)
French mathematician and physicist, he was governor of Lower Egypt during the Napoleonic occupation. He participated in several expeditions and demonstrated an interest in hieroglyphs. His research into heat diffusion and propagation was a highlight of his career.

The Deir el-Medina Settlement

This prosperous Ancient Egyptian town, populated by workmen, was located in a valley between Qurnet Murai Hill and the Theban mountains, opposite what is now Luxor. At the beginning of the twentieth century, archaeological excavations shed light on tombs, houses, and paraphernalia.

Working class neighborhood

Thutmose I reigned in Egypt from approximately 1504 to 1492 BCE, and ordered the construction of his tomb on the hillside, sheltered from pillaging. Workmen and artisans, dedicated to excavating and designing burial chambers, took up residence in the region. A settlement of dozens of houses was created, surrounded by a wall (beyond which houses would later be built). Deir el-Medina remained occupied for almost 500 years.

CONSTRUCTION
Adobe walls were used to build the houses, with the rear wall leaning on the external wall.

WALL
It had at least two gates and measured over 20 ft/6 m in height and 3 ft/1 m wide.

INHABITANTS
In addition to workmen and artisans, scribes and doctors lived in the town.

Ostraca

Fragments of stone or ceramic on which scribes recorded information. Their discovery provided a whole range of information on Deir el-Medina.

MAIN STREET
Crossed the town longitudinally.

Pyramid tombs

Workmen built their own tombs when they were not working on the royal tomb; they were given free time to do this every ten days.

Clay pyramid

Sacred place

Patio

Entrance

Underground chamber and lobby

Burial chamber

TERRACE
A place for socializing on warm nights. Accessible using an internal staircase.

ROOFS
A log structure covered in palm leaves and adobe.

Limestone column

Clay oven

BASEMENT
Excavated from the rock, it was used for storage or a cellar.

KITCHEN
At the far end of the house. Its straw roof allowed smoke to escape.

RECEPTION
With an alcove to honor their ancestors and a false door, which hid an altar.

FIRST ROOM
Accessed by descending three or four steps.

Housing

Simple, with three or four rooms in a row. The type of furniture used was discovered when excavating the tombs.

The Rosetta Stone

The writing system used in Ancient Egypt from 3100 BCE to 400 CE was not deciphered until 1822. The discovery of the Rosetta Stone made it possible to establish the meaning of the hieroglyphs.

The hieroglyphic writing system

A mix of both ideographic and consonant systems. Thus, it comprises four different types of symbols:

Ideograms Symbols that exclusively represent objects in graphic form.

→ r' → Ra = Day/Sun

Phonograms Symbols that represent the pronunciation of a letter.

→ y

Syllabic Symbols that represent the pronunciation of more than one consonant.

→ **nb** → neb
→ **iwn** → iun
→ **rnpt** → renepet

Determinatives Symbols that sometimes serve as markers in words to indicate their semantic role.

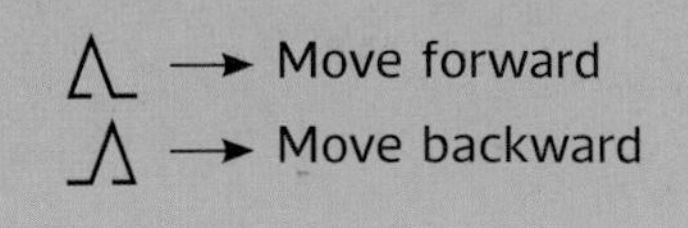

DECREE OF PTOLEMY V
The text written on the stone corresponds to a decree dictated by a council of priests affirming the royal cult of Ptolemy V. It was published in Memphis in 196 BCE.

Three types of writing

On the stone, the text was carved out in three different scripts: hieroglyphs, demotic, and Ancient Greek. Using the latter, researchers were able to decipher the hieroglyphs.

45 in/114 cm

Text in hieroglyphs (script of the gods)

Text in demotic (script of the people)

Text in uncial Greek (script of the government)

GROUPS OF SYMBOLS
Hieroglyphs were not written in a linear sequence, like the letters of the alphabet. They were grouped into imaginary squares or rectangles, to ensure the layout was harmonious.

Expanded text

READING DIRECTION
The Egyptians wrote from left to right as well as from right to left. Thomas Young was the first to establish in which direction hieroglyphs were to be read, based on the direction of the heads of figures contained therein, such as animal heads.

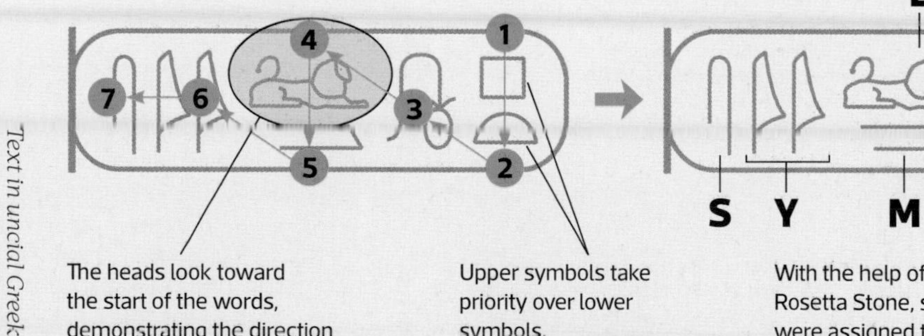

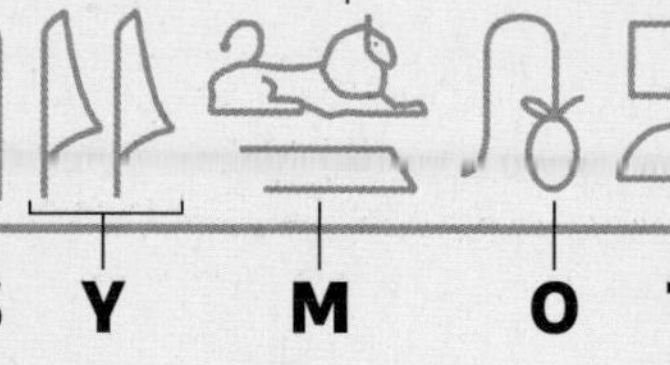

The heads look toward the start of the words, demonstrating the direction in which text should be read.

Upper symbols take priority over lower symbols.

With the help of the Rosetta Stone, sounds were assigned to the symbols.

The names of Pharaohs or queens were encompassed in cartouches like this one.

The discovery

The Rosetta Stone was discovered on July 15, 1799, by Napoleon's troops close to the city of Rashid (Rosetta). Since 1802, it has been displayed at the British Museum.

THOMAS YOUNG
Demonstrated that the hieroglyphs on the Rosetta Stone corresponded to the sounds made when pronouncing the name "Ptolemy."

JEAN-FRANÇOIS CHAMPOLLION
The first to decipher the entire text and the sounds that corresponded to each of the Egyptian hieroglyphs.

The Site at Tanis

Located on the Nile delta, Tanis was the capital of Egypt during the Third Intermediate period. This site, of significant archaeological wealth, houses the royal burial sites of the monarchs from the twenty-first and twenty-second Dynasties.

THE MASK OF PSUSENNES I
The gold burial mask was found over the face of the monarch's mummy, inside several coffins and sarcophagi. Its royal nature can be seen in the use of the royal serpent, the braided beard of the gods, and the wide decorative collar.

The city of Tanis

Tanis reached its peak during the twenty-first Dynasty. Psusennes I founded a great temple dedicated to Amun, the city's deity, along with his wife Mut and son Khonsu. The city was a trading and strategic hub until it was abandoned during the sixth century CE, as a result of fear of a great Nile flood.

The discoverer

Pierre Montet discovered the royal burial site at Tanis on February 27, 1939, and led its excavation until 1956. A passionate photographer, he took thousands of photographs of his discoveries. Despite the importance of his findings, he never received the media support received by Howard Carter with his discovery of the tomb of Tutankhamun.

The burial site

The royal tombs of Tanis house the tombs of Psusennes I and Amenemope (twenty-first Dynasty), of Shoshenq II, Osorkon II, and Shoshenq III (twenty-second Dynasty), in addition to family members and dignitaries. Located in the southeastern corner of the Great Temple at Tanis, most of the tombs were found intact by their discoverer.

THE MYSTERY OF RAMESSES II
In Tanis, statues of Ramesses II were found, along with the remnants of monuments erected by this Pharaoh at Pi-Ramesses, the capital that he founded. This led experts to believe that Tanis was actually Ramesses' capital; however, it was later discovered that it had been transferred stone-by-stone from Pi-Ramesses to Tanis because of a change in the course of the Nile.

TREASURE
The paraphernalia found in the burial chambers comprised valuable objects crafted from gold, silver, and semiprecious stones, such as this golden pectoral of Amenemope (right) and the golden armlet and sandals (above right) of Shoshenq II. The silver sarcophagus of Psusennes I is also particularly noteworthy.

Valley of the Kings

This valley, located on the western bank of the Nile, close to Luxor (ancient Thebes), houses the great necropolis where the Pharaohs of the New Kingdom built their tombs; this included the tomb of Tutankhamun, the most famous of all.

Wall painting from the funerary chamber of Thutmose IV.

Royal tombs

Since the nineteenth century, more than 60 tombs and chambers belonging to the Pharaohs, queens, priests, and other elites from the eighteenth, nineteenth, and twentieth Dynasties have been found. Tutankhamun's tomb (KV62) was the last to be officially discovered in the valley. In 2005, KV63 was found, along with a storage chamber, and in 2008 two new chambers (KV64 and KV65) were found. However, their contents remain unknown to this day.

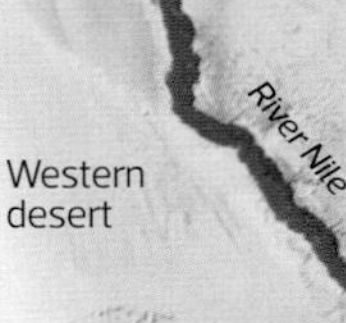

Theodore M. Davis

1837–1915

American archaeologist known for his excavation work in the Valley of the Kings between 1902 and 1914. Under his instruction 30 tombs were discovered, and he was just 6 ft 6 in/2 m from the entrance of the tomb of Tutankhamun, although he only found objects linked to him.

The Valley of the Kings in 1905.

TOMB KV5

Belonging to the children of Ramesses II, it is the largest of the entire valley. Discovered in 1825 by James Burton and examined in 1902 by Howard Carter, excavation did not begin until 1987. In 1995, they discovered passages and several chambers beneath the rock. To date, 121 rooms and corridors have been discovered. However, it is believed more are hiding in wait.

One of the corridors, with a statue of Osiris at the end.

Looting

To prevent looters entering the royal tombs, the Pharaohs abandoned the idea of constructing great pyramids and started to build their tombs underground, with entrances hidden among the rocks and sand. Nonetheless, almost all the tombs discovered in the valley were looted before the end of the twentieth Dynasty.

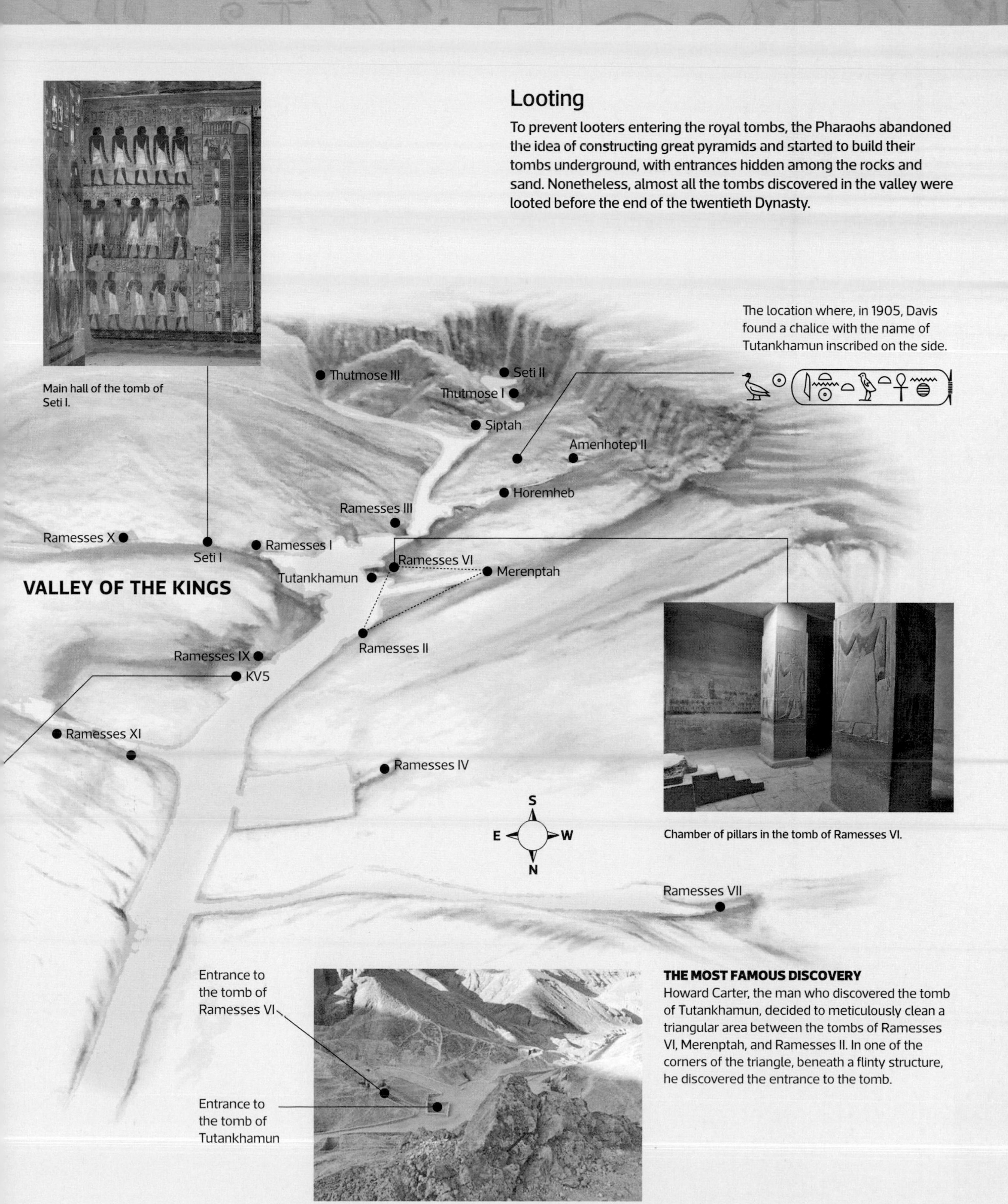

Main hall of the tomb of Seti I.

Chamber of pillars in the tomb of Ramesses VI.

THE MOST FAMOUS DISCOVERY

Howard Carter, the man who discovered the tomb of Tutankhamun, decided to meticulously clean a triangular area between the tombs of Ramesses VI, Merenptah, and Ramesses II. In one of the corners of the triangle, beneath a flinty structure, he discovered the entrance to the tomb.

Howard Carter

By 1914, the Valley of the Kings had already been subjected to significant excavation work, although Carter believed that the biggest find was yet to be uncovered. He did not stop until he discovered the tomb of Tutankhamun, a find that led to his worldwide acclaim.

A passion for Egypt

This British Egyptologist was born on May 6, 1873, and inherited his talent for illustration from his father; from a very young age, he demonstrated an interest in Egypt. Due to health problems, he attended school infrequently and was given an artistic education by his father. He traveled to Egypt as an artist for the first time at the age of 17. From then on, he spent most of his career in that country, until he found the tomb of Tutankhamun in 1922. He died in England on March 2, 1939.

STRONG CHARACTER
Carter was a solitary, brash, and stubborn man, not known for his diplomatic skills. A bad-tempered outburst cost him his job in 1905. However, his tenacity led him to the most important treasure find in the history of Egyptology.

His career

1892
Worked in Egypt for the first time as a restorer in Deir el-Bahri (Luxor), under the guidance of the prestigious Egyptologist Flinders Petrie.

1899
Named Chief Inspector of Antiquities in Upper Egypt and, later, Lower Egypt. He remained in this position until 1905, when he was removed from his post following a confrontation with French tourists.

1907
He met Lord Carnarvon and a great friendship was born. In 1908, Lord Carnarvon financed Carter's excavations in Thebes.

1914
Carter and Carnarvon obtained excavation rights for the Valley of the Kings.

1922
In November, he uncovered the steps leading to the tomb of Tutankhamun.

1923
On February 16, he opened the funerary chamber of the Pharaoh.

AT THE TOMB OF TUTANKHAMEN: THE OPENING CELEBRATIONS.

Le Petit Journal
illustré

ABONNEMENTS

Dans la poussière des Tombeaux

Front page news. The world's press spread news of the discovery and the Valley of the Kings became a pilgrimage site.

THE MASK
When Carter opened the last sarcophagus, he found the golden mask of Tutankhamun, one of the most beautiful pieces of the Pharaoh's treasure.

The discovery

After eight years of work to find the tomb of Tutankhamun, Lord Carnarvon, sponsor of the excavation, informed Carter that the time had come to abandon the project. Carter begged him for an extra year of breathing space and, on November 4, 1922, beneath the entrance to the tomb of Ramesses VI, a step carved into the rock appeared. It was the first of 12 steps that would lead to the tomb of Tutankhamun. Behind the sealed door, distributed among different chambers in the tomb, great treasure lay in wait.

Carter and Lord Carnarvon. This image shows them tearing down the wall of the funerary chamber.

Carter and A. R. Callender. Opening the gilded wood sarcophagus that contained the body of Tutankhamun.

The Tomb of Tutankhamun

It is relatively small for the Pharaoh's final resting-place—a sign that it was not made with the young king in mind. Its size contrasts with the wealth of treasure found in the four chambers within the tomb.

ANNEX
The entrance to the annex was hidden behind the furnishings. It was the last to be examined, since there was a wide range of artifacts piled up.

The structure

The tomb follows the structure of others found in the Valley of the Kings: a passage that leads to an antechamber, which in turn gives way to the funerary chamber that housed the Pharaoh's sarcophagus. This chamber is hidden behind a sealed wall, and the entrance was guarded by two life-size statues of Tutankhamun. Two side rooms were identified as the "Annex" and the "Treasury."

ANTECHAMBER
The entire room was sealed off. When Carter broke through the first door, he found a room full of the Pharaoh's objects, many crafted from carved gilded wood or gold itself.

THE CORRIDOR
A sloping corridor measuring 6 ft 6 in/2 m high and 5 ft 7 in/1.7 m wide led to the antechamber entrance.

ENTRANCE
In November 1922, after eight years of searching, English archaeologist Howard Carter found the entrance to the tomb of Tutankhamun, hidden in the rocky floor of the Valley of the Kings.

THE LOST TOMB
Two centuries after the tomb of Tutankhamun was built, the Egyptians excavated the sepulchre of Ramesses VI, thereupon effectively burying Tutankhamun's tomb.

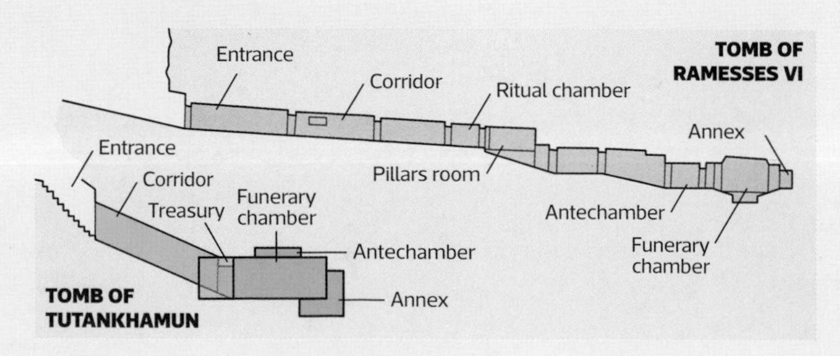

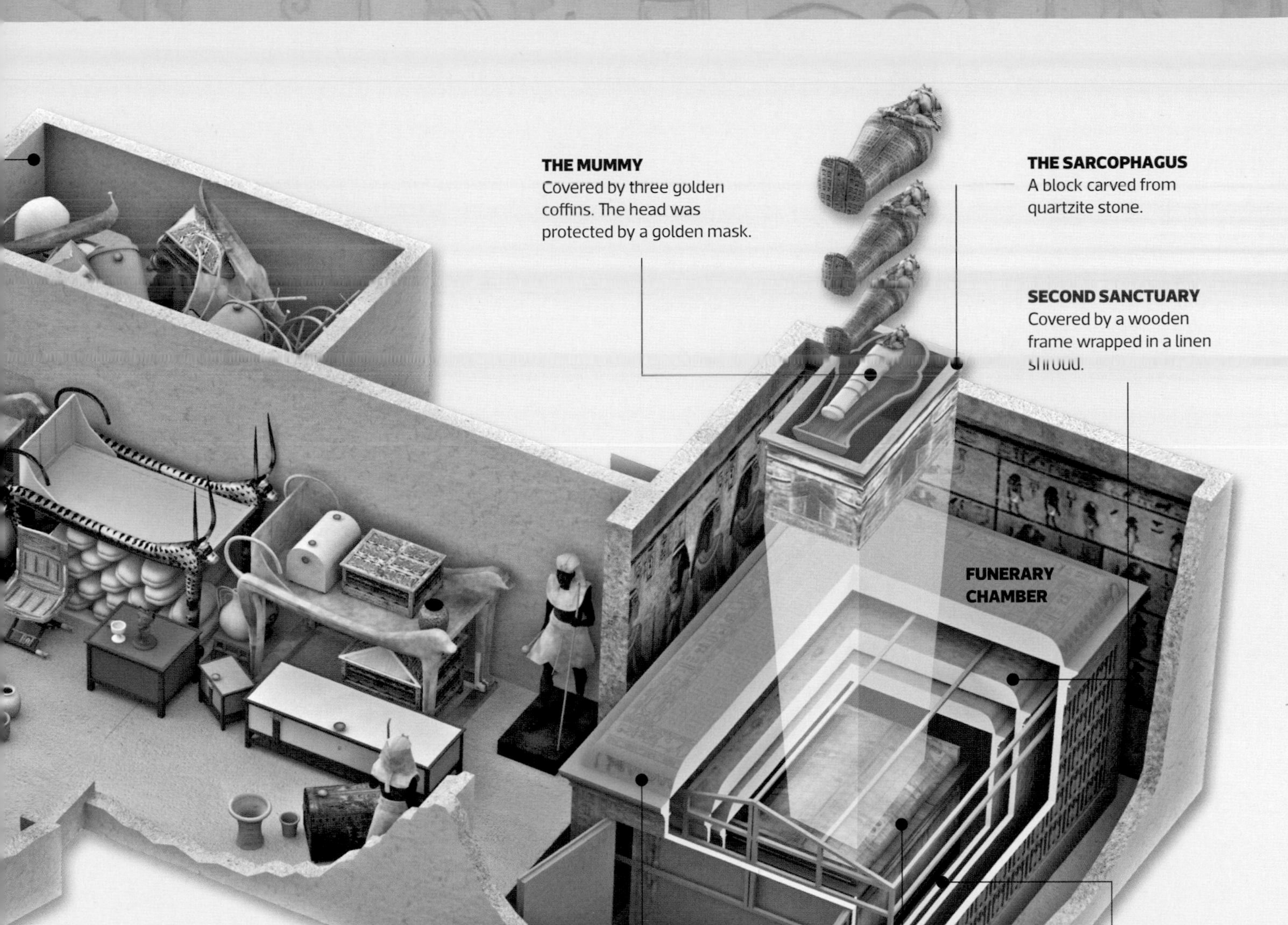

THE MUMMY
Covered by three golden coffins. The head was protected by a golden mask.

THE SARCOPHAGUS
A block carved from quartzite stone.

SECOND SANCTUARY
Covered by a wooden frame wrapped in a linen shroud.

FIRST SANCTUARY
The first sepulchre was made from carved cedar wood with fragments of blue earthenware.

THE TREASURE
Chamber located behind an open door guarded by a statue of Anubis. It contained the canopic chest.

CANOPIC CHEST
Protected by four gods, it held the liver, lungs, stomach, and intestines of the Pharaoh.

THIRD SANCTUARY
Gilded and, like the others, engraved with religious inscriptions.

FOURTH SANCTUARY
Carved with images of the gods. Isis and Nephthys guarded the door, while Nut and Horus guarded the roof.

The Pharaoh's Treasures

One of the reasons that the discovery of the tomb of Tutankhamun is considered so special is that the funerary paraphernalia was found intact. The objects discovered in the different chambers are of unimaginable archaeological value.

SYMBOLS
The monarch's effigy was adorned on the forehead with a vulture and a cobra, the emblematic gods of Upper and Lower Egypt, respectively.

The mask

The mummy of the Pharaoh was found covered with a skillfully produced mask of pure gold, inlaid with blue glass and semiprecious stones; it is currently housed in the Cairo Museum. It measures 21 in/54 cm in height and weighs just over 22 lb/10 kg. Wearing the royal striped headdress and adorned with a thin braided beard, Tutankhamun is presented in the form of the god of the dead, Osiris. Although the mask is the most spectacular piece, it is just one of 150 found covering the Pharaoh's body.

FAN
Eight fans were found. The one in the image is of ebony thickly covered with gold, decorated with inlaid glass and calcite at the center—the cartouches containing the names of Tutankhamun are particularly noteworthy. The feathers did not withstand the test of time, although astonishingly, those on another fan were preserved intact.

GOLDEN THRONE

Six seats were found. The most impressive was this very elaborate golden throne that featured arms and was covered in gold and silver sheets. The legs replicate those of a lion, while the side panels take the shape of winged cobras. A scene depicting the Pharaoh and his wife is illustrated on the back panel.

CROWN

The ornately decorated royal diadem was found on the mummy's head. It is made of gold with stone inlays. The vulture and cobra adornments were detachable, and were found beneath the body.

JARS

Eighty jars were found in the tomb. Their strange shapes amazed Carter. The one shown above is a chalice made of calcite in the shape of a lotus flower.

STATUETTES

These small wooden sculptures covered in gold, part of a series of 32, represent the Pharaoh. They measure 33.5 in/85 cm in height and were of significant ritualistic value.

ANUBIS

Tutankhamun in the form of the god Anubis, responsible for the embalming process. The figure, carved from wood varnished with black resin, rests on a shrine containing ritual objects.

An Overview of Tutankhamun

Led by Egyptologist Zahi Hawass, a team of archaeologists examined a scan of the mummy, which allowed them to gain a better idea of the physical features of the Pharaoh. As such, it was possible to clarify queries about his mysterious death and gain more information about certain details of his life.

Advanced technology

Tomography, in use since the 1930s, is a diagnostic system that takes various X-ray images and provides a cross-section of a given part of the body. In 1972, a scanner was invented to perform this task digitally and process results using a computer. In 1996, a volume-generation technique was invented to obtain images in 3-D. This technology was used at the beginning of 2005 to examine the mummy of Tutankhamun.

A DETAILED EXAMINATION OF THE SKULL
The king's head was scanned in sections of just 0.02 in/0.62 mm long, to highlight its complex structures to the greatest level of detail possible.

Cause of death

The circumstances of the young Pharaoh's death have been the cause of much speculation. The most recent scientific analysis suggests that malaria, along with complications resulting from a hip fracture, caused or at least contributed to the premature death of Tutankhamun at the age of 19.

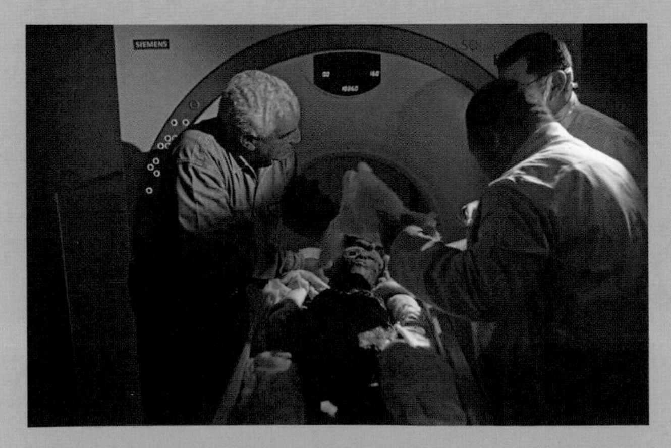

Analysis. Tomography has made it possible to understand the deformities and illnesses suffered by the young king.

SKIN COLOR
The Pharaoh's exact skin color will probably never be known. Restorers have based reconstructions on paintings and busts of Tutankhamun (right), in addition to those of his close family members. As a reference, variations in the skin color of the present-day population of Egypt were used, with an intermediate tone employed.

Art and science

Élisabeth Daynès, an artist specializing in reconstructing animals and people from ancient times, was responsible for providing the Pharaoh with the most precise likeness ever achieved, using data from the tomography.

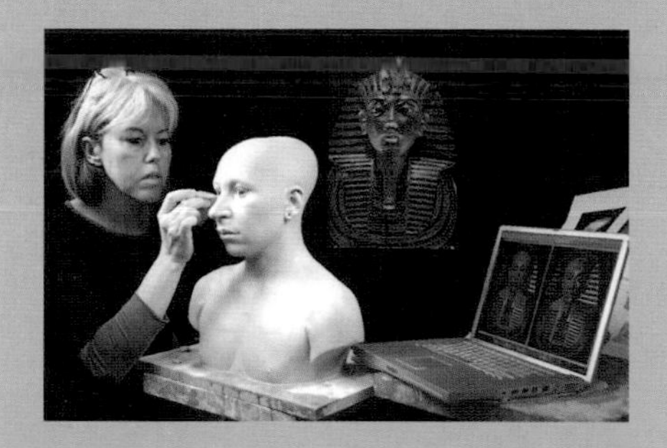

How the reconstruction was done

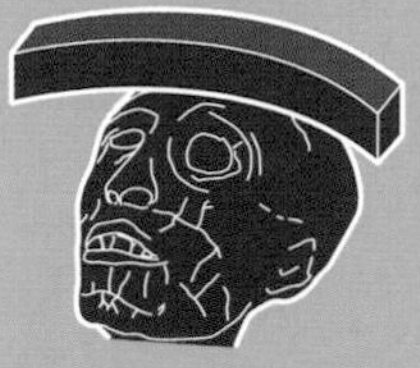

SCAN
The scanner took almost 1,700 digital X-ray images of the mummy, which were loaded onto a computer.

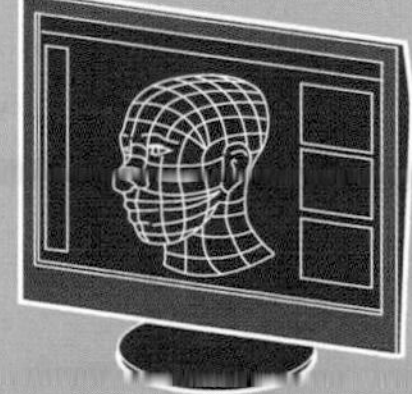

3-D MODEL
Software was used to create a volumetric projection, to obtain a three-dimensional image.

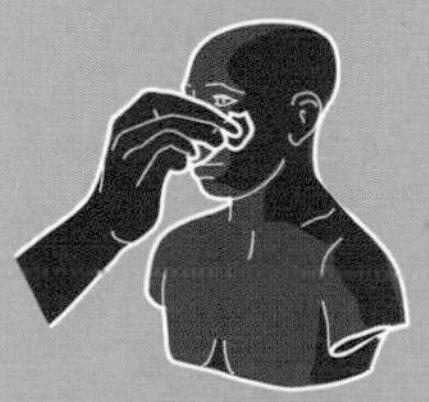

RECONSTRUCTION
Using the 3-D image, forensic anthropologists worked on a real skull cast, reconstructing the physical features of the Pharaoh.

THE FACE OF THE KING
Using sculptures of the Pharaoh and his family, a model was created that offered an adjusted reconstruction of the face of Tutankhamun at the time of his death.

Identifying Mummies

Establishing the identity of a person who died thousands of years ago is the result of a multidisciplinary process that combines information from historic tests, archaeological findings, anthropology, and technological advances in medicine.

THE AGE OF AKHENATEN
A tomography of the alleged skeleton of Akhenaten revealed that he suffered from osteoarthritis, an illness that appears in later life, disproving the belief that he had died at the age of 25.

X-rays

The use of X-rays to identify mummies has taken a backseat since the appearance of computerized tomography. However, the technique is still used in the first instance and is especially useful in locating bone injuries or skeletal disorders.

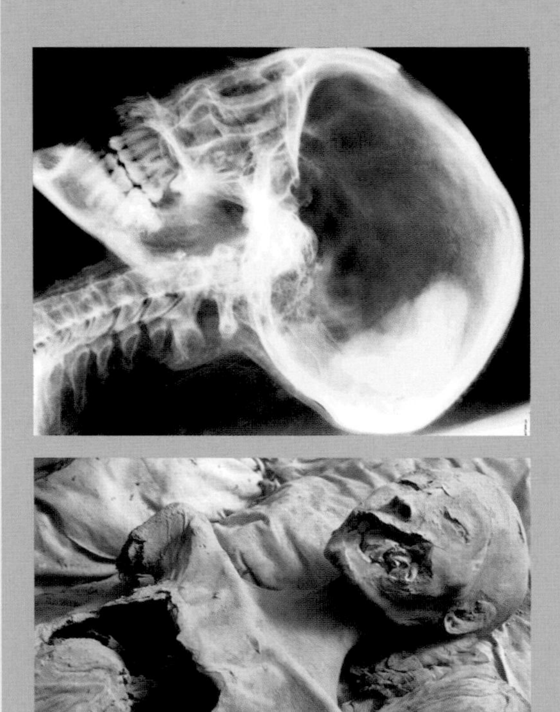

IDENTIFICATION USING TEETH
Scan of the skull (top) and the mummy of "The Younger Lady," found in KV35. The X-ray films taken of the skull serve to establish the age, using the mummy's teeth and the joints of the skull. In this instance, it was established that "The Younger Lady" was aged around 25 when she died.

Tomography

This X-ray system provides 3-D images in high resolution. It is possible to detect anomalies with great precision, making it possible to identify the person in question rather than having to explore their biographical data.

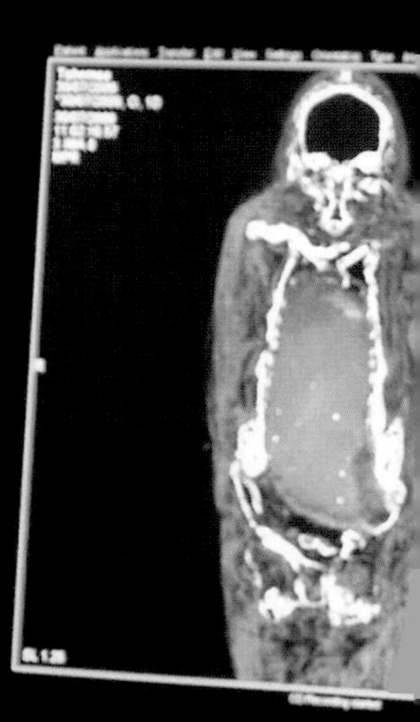

DISCOVERING PARENTAGE

After two years of DNA analysis and tomographies taken on Tutankhamun and 15 other mummies, the results of an investigation into the mummy's family were published in February 2010. They concluded that Akhenaten was his father, the woman known as "The Younger Lady" was his mother, and Queen Tiye was his grandmother. However, in 2013, new interpretations carried out by Marc Gabolde opened the door to the theory that the mother of Tutankhamun was Nefertiti, wife and first cousin of Akhenaten.

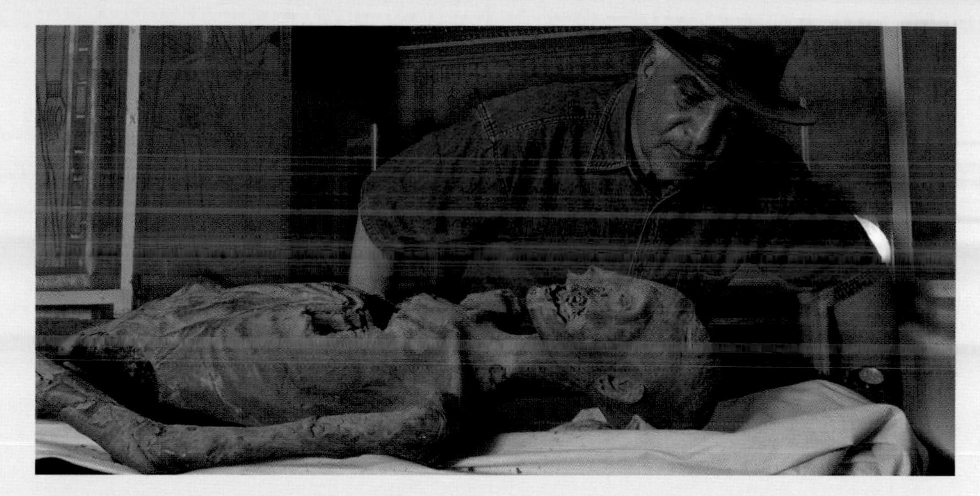

Egyptologist Zahi Hawass with the mummy known as "The Younger Lady," possibly the mother of Tutankhamun.

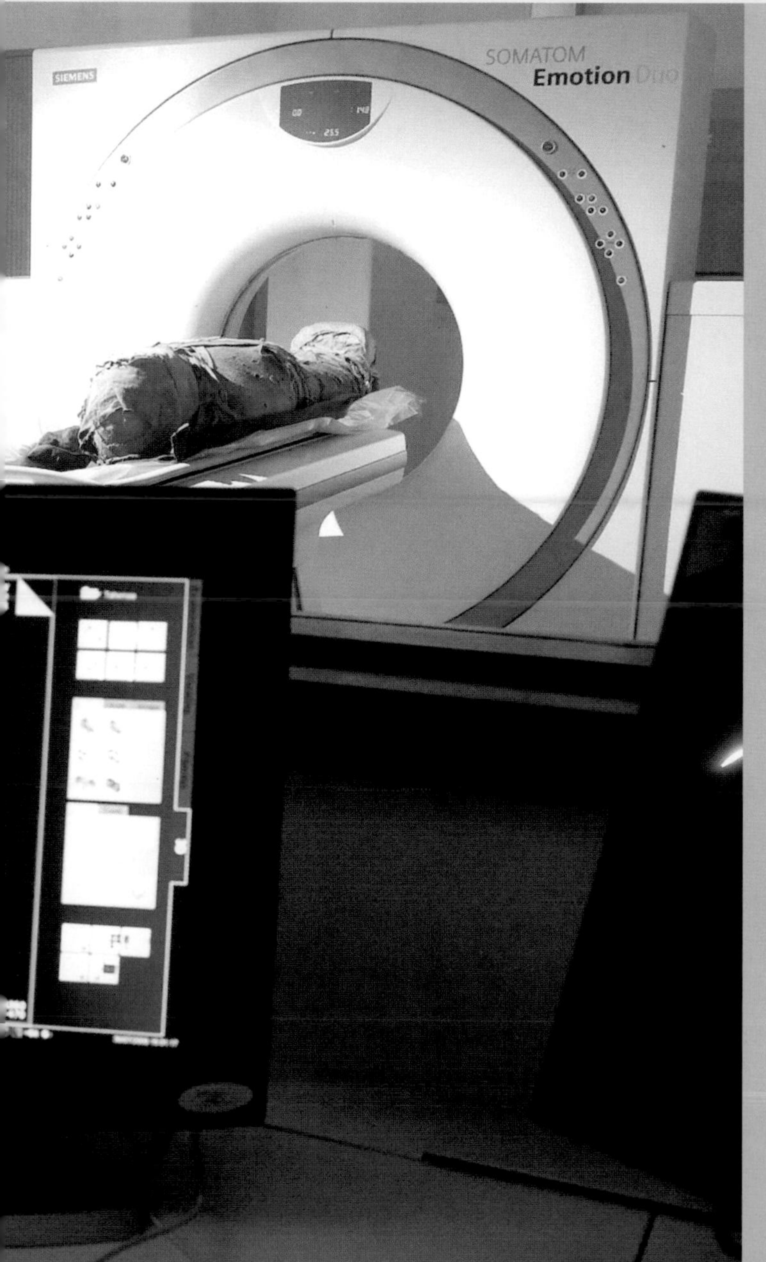

DNA analysis

DNA identification tests have not only made it possible to discover the identity of unknown mummies, they have also served to confirm the practice of royal incest during the eighteenth Dynasty.

1. Uncontaminated

Initially, geneticists take the necessary precautions to ensure the mummy is not contaminated with traces of their own DNA during the process.

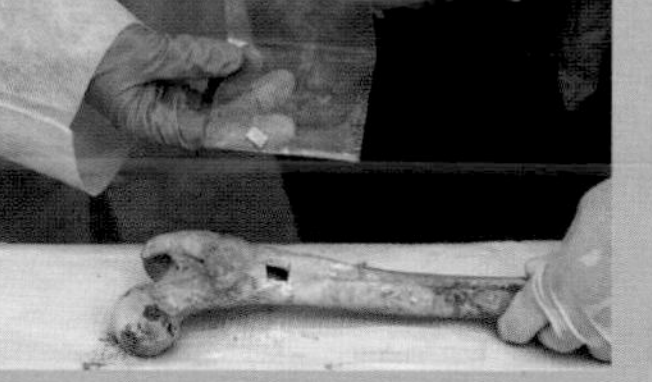

2. Sample taking

They extract tissue samples from various parts of the mummy, from inside the bones in particular, as this area is most protected from external contamination.

3. Analysis

In the laboratory, the genetic composition of the mummy is analyzed using the tissue sample. Chromosomes are then isolated, establishing the gender and location of gene-sequence segments.

4. Matches

The results are compared to probable parents. If there are matches in at least eight segments, the paternal or maternal relationship can be confirmed.

Robotic Explorers

The pyramid of Khufu (Cheops in Greek) has four narrow ducts, the purpose of which remains a mystery. Two of them begin in the king's chamber and end outside, while another two start in the queen's chamber, although their destination is unknown. To explore them, robots equipped with cameras were built.

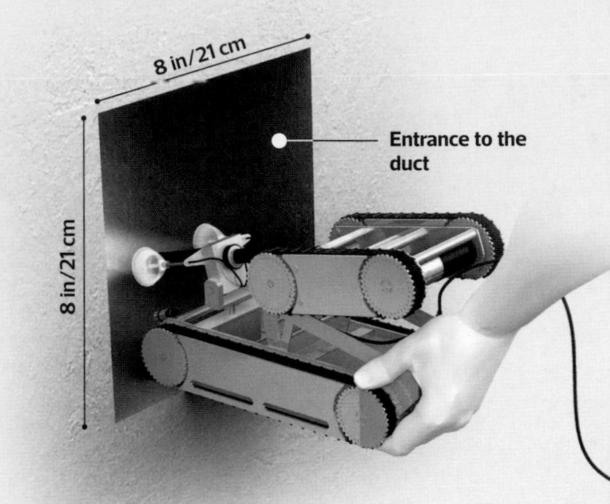

Upuaut robots

Designed by German engineer Rudolf Gantenbrink in 1992, the Upuaut were used to explore the upper ducts and establish that they ended outside. The first robot was replaced by an improved model, Upuaut-2, which traveled the lower ducts, in which it discovered slabs that blocked its path.

Halogen lights

The ducts

The two upper channels were discovered in the seventeenth century. The two lower ducts were discovered by British engineer Waynman Dixon in 1872.

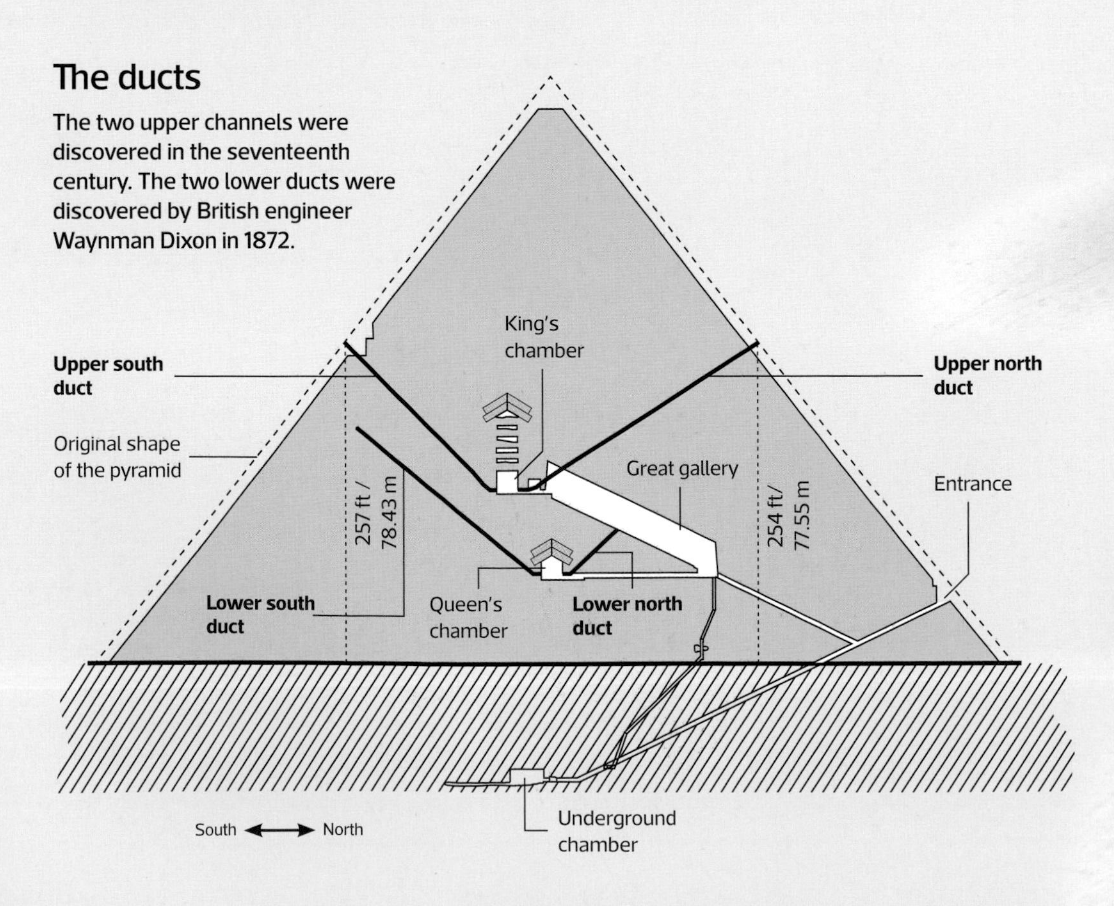

WHAT WAS THE PURPOSE OF THE DUCTS?

Investigators hired by the government were unable to answer this question. What they were able to establish was that they were not drilled after the pyramid was completed, but were planned in advance.

Upuaut-2 Robot

STRUCTURE
To create the parts of the robot's structure, aluminum, similar to the type used on aircraft, was cast and molded.

Electric motors

IMAGING
Miniature video camera with a CCD recording system. It can turn left and right on its axis and freeze the image.

POWER SUPPLY
The electrical current to drive the robot was supplied using an umbilical cable of 0.16 in/ 4.2 mm in diameter.

14.6 in/37 cm

4.7 in/12 cm

TRACK
Rubber track to gain better grip.

LASER GUIDANCE SYSTEM
Serves to measure the inside of the tunnel.

PROPULSION
Seven independent electric motors controlled the upper and lower wheel system, providing 88 lb/ 40 kg of thrust (in ideal driving conditions).

MAXIMUM AND MINIMUM HEIGHT OF THE ROBOT

4.7 in/ 12 cm

11 in/ 28 cm

Other Egyptologists

From the Napoleonic Campaign to modern times, numerous academics have dedicated their lives to the study of Ancient Egypt. Some of them form part of the history of Egyptology, whether as a result of their discoveries, their methods, or the fact that they broke with tradition at the time.

Giovanni Belzoni

1778–1823

An adventurer born in Padua. He lived in Egypt between 1815 and 1819, where he participated in several excavations. He discovered six tombs at the Valley of the Kings, entering the larger Abu Simbel temple for the first time and entering and reaching the funeral chamber in the pyramid of Khafre (Chephren in Greek). Many of his findings form part of the collection of the British Museum in London.

Karl Richard Lepsius

1810–84

Considered one of the main Egyptologists after Champollion, he studied Egyptian collections in Europe and completed research started by Champollion on Egyptian grammar. In 1842, he participated in a scientific expedition to Egypt and Sudan, discovering the Decree of Canopus in 1866. His discoveries form a significant part of the collection of the Egyptian Museum in Berlin.

Ippolito Rosellini

1800–43

Friend and student of Champollion, they participated in a scientific expedition to Egypt together to copy hieroglyphs and excavate, with a view to increasing Egyptian collections in France and Italy. On his return, and after the sudden death of Champollion, he worked alone to classify the Egyptian findings.

EGYPTIAN AND NUBIAN MONUMENTS

Page featuring in the work of Rosellini's *I Monumenti dell' Egitto e della Nubia*, comprising ten volumes. The first volume was published in 1832. In total, it contained 390 extremely useful illustrations.

Auguste Mariette

1821–81

Scholar of Ancient Egypt and hieroglyphs. He worked at the Louvre and participated in various Egyptian excavations. One of his most noteworthy findings was the Serapeum at Memphis, in 1851.

Facade. Current facade of the Cairo Museum.

THE FIGHT AGAINST LOOTING

To Mariette, after so many years of discovery and research on Ancient Egypt, preserving the material found was of the utmost importance. In order to stop looting and ensure the preservation and study of Egyptian heritage, he suggested the creation of the Boulaq Museum, which later became the Cairo Museum.

William Matthew Flinders Petrie

1853–1942

A self-educated English Egyptologist who participated in 38 excavations in various places in Egypt over a period of more than 40 years. During this time, he employed revolutionary methods based on observing all the objects found and studying the materials. His main professional objective was that his publications, totaling more than 100, would serve as a valid and indisputable source of information for many years.

Zahi Hawass

b. 1947

Egyptian Egyptologist interested in recovering and preserving heritage. He was head of the Supreme Council of Antiquities in Egypt for almost ten years, and also the Egyptian Minister of Antiquities. He has led numerous excavations and is famous for studying mummies using computerized tomography, in particular Tutankhamun and a number of his relatives.

CHAPTER

HISTORY AND ORGANIZATION OF THE STATE

CONSTRUCTION OF AN EMPIRE

Over time, the civilization that was born on the banks of the Nile became a huge empire governed by the all-powerful Pharaohs. Maintaining territorial and social unity was the principal challenge facing the numerous dynasties that ascended the Egyptian throne.

As asserted by Greek historian Herodotus, "Egypt is the gift of the Nile." Between 13000 and 10000 BCE, the valley through which this river passes was irrigated by abundant rainfall. The flow of the river increased significantly and the number of pastures multiplied. This transformation, brought about by climatic change, resulted in the presence of new animal species, such as various breeds of wild donkey, as well as crops such as millet, sorghum, and African rice. Groups of hunter-gatherers also flourished in this new ecosystem, discovering the art of domesticating plants and animals. The development of agriculture and livestock resulted in a sedentary lifestyle replacing the previously commonplace nomadic existence. Here, during the Neolithic period, among the small agrarian settlements that shaped the Nile Valley, the Ancient Egyptian civilization was born.

The birth of a great civilization

Toward the end of June, following the strong rainfall during the monsoon in Abyssinia, the river swelled, acquiring a green color due to the amount of vegetation swept away by its waters. As the ice melted on the mountaintops, the river level rose; abundant in reddish clay, the rising waters resulted in great floods. As the water subsided, the soil was left rich in silt, the best fertilizer that agriculture could have hoped for. The valley's ancient inhabitants also learned to tame the flooding of the river, developing irrigation systems that made agrarian development possible. The more advanced settlements experienced production surpluses, promoting trade and, at the same time, hegemony over less advanced settlements. Based on archaeological excavations, it is known that between 4000 and 3000 BCE, the enclaves of El-Badari, Faiyum, El-Amra,

AKHENATEN Bust of Amenhotep IV, Pharaoh of the eighteenth Dynasty, who later changed his name to Akhenaten.

BLUE, THE COLOR OF THE NILE
Earthenware statuette dating back to 1900 BCE. The color blue, associated with the River Nile, was used to give color to the animals that lived in its waters and on its banks.

Nagada, El-Kab, and Gerzeh, among others, gained their own particular profile. Objects excavated demonstrate a significant level of development in the tools and weapons used, and a growth in religious beliefs around the cult of Hathor, the goddess of fertility and reproduction.

Around 3000 BCE, economic development resulted in urban revolution: the most powerful villages became cities, in which different elementary forms of social control became apparent, i.e. the state. Palaces, burial sites for the royal family, temples, accommodation for the military and priests, and trade centers were erected. Among these cities, Nekhen is worth particular mention; located in Upper Egypt, it was mentioned by Herodotus, and would soon, in partnership with other cities, unify the lands across its territory.

Social organization

The basic features of Egyptian history can all be traced to this period. The center of power was transferred from Nekhen to Memphis, where the Nile breaks up into the delta. The government assumed a distinctly military and religious character, the pinnacle of which was occupied by the Pharaoh.

The social and political organization of Ancient Egypt took shape at this time. Theocratic in nature, it was based on the relationship between religion and basic subsistence, which was dependent on the Nile. The unity of faith and daily life preceded the concentration of all powers, both military and spiritual, in a single individual: the Pharaoh, a name that translates as "great house." As such, he occupied the highest position of the government and the religious hierarchy and was worshipped as the supreme god of a diverse pantheon. The Pharaoh entrusted state administrative responsibilities to the vizier, similar to a prime minister, who would become more important in later periods, and the nomarchs, or provincial chiefs. The second step of the pyramid was occupied by the priestly caste, followed by administrative workers, among whom the scribes were particularly important; they were responsible for setting down

CERAMICS
Amphora decorated with riparian motifs typical of the Upper Nile, dated to 3500–3100 BCE.

in writing imperial edicts and laws, reports, trade agreements, and sacred texts. In fourth place was the military caste and, in fifth, traders and artisans. Peasants occupied the sixth social caste, and beneath them came the lowest stratum—slaves.

The Old Kingdom

The main policy was to maintain the unity of Upper and Lower Egypt, ward off encroachments from nomadic tribes, and keep the gold and precious stone mines active. This policy was consolidated by the assimilation of symbolism from both Upper and Lower Egypt, such as the red crown of the South and the white crown of the North.

It appears that in Lower Egypt there were constant pockets of resistance to unification. This was most apparent during the reign of Anedjib, who was forced to face fierce rebellions. Under the rule of Khasekhem, this confrontation became a religious war between the followers of Horus and those of Set. Finally, Khasekhem was victorious, and his success was confirmed by a change in his name: Khasekhem, which meant "The Sole Powerful One," was changed to Khasekhemwy, which meant "The Two Powerful Ones."

Around 2200 BCE, during the reign of Pepi II, the political situation deteriorated significantly: harsh droughts and a drop in the level of the Nile resulted in hunger that led to popular uprisings. The invasion of tribes from Asia hastened the end of the Memphis-based monarchy. Akhtoy, known as Khety I, the monarch of the city of Herakleopolis, led a coup d'état and deposed Neferkara I, the last king of Memphis. Around 2030 BCE, Mentuhotep II of Thebes conquered Herakleopolis and unified Egypt under his control. This would lead to the beginning of the so-called Middle Kingdom.

Around 3000 BCE, economic development resulted in urban revolution: the most powerful villages grew rapidly to become cities

CARVED SPOON
This utensil was used for cosmetics. The female figure holds a container of ointment, supported on her shoulder. Piece from the eighteenth Dynasty.

CROCODILE
Figure of a crocodile made from bronze, found in Faiyum and dated to around 1850 BCE, during the twelfth Dynasty (Middle Kingdom).

THE PHARAOH QUEEN
Sculpture of Queen Hatshepsut (eighteenth Dynasty) in the shape of a sphinx. Daughter of Thutmose I, she ruled Egypt from 1479 to 1458 BCE.

The Middle Kingdom

Mentuhotep II strengthened the army and restructured the administrative system. He bolstered the role of the vizier and appointed governors, directly subject to his control, in the different districts of the kingdom. Furthermore, he installed the Council of Greats, the role of which was to assess his governance. He proclaimed himself the son of Re, god of the Sun, and the supreme deity of the pantheon. Abroad, he focused on his military campaigns against Nubia, reestablished trade routes, and restarted mining.

However, conflicts between his heirs weakened the new monarchy, and as a result, in 1981 BCE, the vizier Amenemhat ascended the throne. Twenty years into his reign, in order to avoid traditional conflicts about succession, Amenemhat I named Senusret I, head of the troops that fought in Nubia, as his successor. Once in power, the latter launched a military campaign against the Kush region and took over the gold, copper, alabaster, and diorite mines. This source of wealth helped the expansion of agriculture in the region of the present-day Faiyum Oasis, located to the west of the Nile on one of its tributaries. Significant development in this region, which became the empire's breadbasket, resulted in Amenemhat III building his funerary complex there, erecting two huge statues at the entrance to the channel uniting the lake and the river. During this period, Egypt expanded its trade route into the Red Sea, where its main client was the Land of Punt (present-day Somalia), a producer of incense. Egypt also maintained significant trade ties with the island of Crete, a naval power during the period, and the city of Byblos, today part of Lebanon, which at the time was the main supplier of wood in the eastern Mediterranean.

Between 1800 and 1500 BCE, various bordering nomadic towns, in particular from Libya and Asia, tried to expand into Egypt. Most came in the form of the Hyksos invasions. Around 1720 BCE, during the reign of Sebekhotep IV, the Hyksos, led by Salitis, occupied Avaris, just a few miles from Qantir (present-day Tell el-Dab'a). After expanding throughout the eastern delta, they took

Ramesses II (1279–1213 BCE) consolidated the new period of Egyptian expansion. The temples at Luxor and Karnak are a reminder of this age of imperial splendor

SENUSRET I
Statuette of Senusret I, who ruled Egypt from 1961 to 1917 BCE, wearing a white crown made from painted cedarwood.

control of Memphis. This situation was reversed when Ahmose I, founder of the eighteenth Dynasty, expelled the Hyksos from the delta, restoring Egyptian control and forming the New Kingdom.

The New Kingdom

As Pharaoh between 1550 and 1525 BCE, Ahmose I advanced on the present-day Gaza Strip and, simultaneously, on Nubia. His son Amenhotep I, who ruled between 1525 and 1504 BCE, extended the kingdom's borders to Canaan and Syria. His successor, Thutmose I, who occupied the throne from 1504 to 1492 BCE, swept aside the small kingdoms of Syria and Palestine, reaching Mesopotamia.

The legitimacy of the Pharaoh was cemented by his marriage to his sister. Thus, they sought to keep the descendants of the dynasty's founder pure. Following the death of Thutmose II, who left no legitimate male descendant, Hatshepsut, his half sister and royal wife, ascended the throne. With the support of the clergy of Amun, Hatshepsut was declared Pharaoh in 1473 BCE, displacing Thutmose III, the son of Thutmose II, and one of his secondary wives. During her reign, Egypt maintained its influence in Asia but failed to make any further progress. Thutmose III only came to power following the death of Hatshepsut. His military campaigns, which involved a formidable cavalry and squadrons featuring chariots of war, in addition to an infantry, were successful, and he took control of Cyprus, Crete, and Babylonia. He erected temples in Karnak, Heliopolis, and Memphis. Thutmose IV did the same, building temples dedicated to the sun cult of Amun.

As a result of the political and religious conflicts, which constantly challenged the central power with numerous local interests, between 1352 and 1336 BCE, Amenhotep IV replaced the cult of Amun with the cult of Aten. He also changed his name to Akhenaten, "servant of Aten," and the name of the capital city Thebes to Akhetaten, "Horizon of Aten," before eventually moving his capital to a new settlement at Tell el-Amarna, at the heart of the kingdom. This political and religious measure deepened the conflict with the priestly caste, which followed the cult of Amun and preached it to the people. Amunist repression united the clergy and nomadic Hyksos tribes, whose belligerence was driven by a desire to control trade routes; furthermore, they regularly attempted to invade the Nile Valley, the heart of Egypt. The Hittite invasion, from the Kingdom of Hatti in central Anatolia, presented a genuine threat to the kingdom's integrity. Tutankhaten, who succeeded Akhenaten, restored the cult of Amun, changing his name to Tutankhamun. Thus, Egypt embarked upon a new period of stability.

This situation culminated in the coronation of the Ramesses Dynasty, whose members, such as Ramesses I, a former military vizier, took advantage of the Assyrian attack on the Kingdom of Hatti to recover territory lost to the Hittites, in particular the Kadesh region. Ramesses II, grandson of Ramesses I, who ruled for 66 years between 1279 and 1213 BCE, ended this new period of Egyptian expansion. The temples erected at Luxor and Karnak, such as the one at Abu Simbel, are a reminder of this age of imperial splendor.

The "dark times"

During the thirteenth, twelfth, and eleventh centuries BCE, the region of Asia Minor and the Mediterranean basin was disrupted by fierce conflicts. Trade relations between Egypt and the Kingdom of Hatti were changed when Assyria conquered the Hittite mining operations. Hatti occupied Cyprus in search of copper, an essential material, together with tin, in the production of bronze. This event involved the displacement of the Achaean kingdoms, the Peloponnesian ancestors of future Greece. Piracy and looting of coastal cities made traditional trade routes a thing of the past.

It fell to Merenptah (r. 1213–1203 BCE), son and successor of Ramesses II, to overcome these "dark times." Over the course of the following decade, the

PTOLEMAIC PERIOD
Columns of the temples of Sobek and Horus at Kom Ombo, built during the reign of the Ptolemaic Dynasty in the second century BCE.

conflicts grew more intense. The "War of the Impure," as the Thebans called it, was waged. This conflict represented a holy war between rebellious followers of Set, with their center in Heliopolis, and the Theban clerics, worshippers of Amun. This battle would result in the division of the kingdom: Smendes declared himself Pharaoh of Lower Egypt, and Herihor, a Libyan mercenary general, declared himself Pharaoh of the remainder of the kingdom's territory.

The rivalry between the two monarchs resulted in the foundation of other centers of power. The city of Herakleopolis grew in strength around the beginning of the delta. Hermopolis, allied with Libyan tribes, assumed control of the central region of Egypt, and, in 725 BCE, Letopolis, located in Sais, the westernmost island on the delta, declared independence. The fate of Ancient Egypt was sealed. In 716 BCE, Piye, a Kushite (Nubian) king, installed a princess of the family in Thebes and proclaimed her God's Wife of Amun. The new Kushite dynasty reestablished Memphis as the capital; nonetheless, attempts to expand into Palestine clashed with Assyrian interests. Led by kings Tiglath-Pileser, Salmanasar V, Sargon, and Sennacherib, Assyria had taken control of Judaea and Israel, and was advancing toward the Nile. It was King Ashurbanipal who brought Thebes under Assyrian control. Over time, they were conquered by Greeks, Babylonians, Persians, and Macedonians, sealing the downfall of the great Egyptian civilization.

Even so, the splendor of the empire lasted centuries, assimilated and imitated by new governors from other locations. An example of this can be seen in the Hellenistic period, inaugurated by Ptolemy I Soter, founder of the Ptolemaic Dynasty in 304 BCE. The Ptolemies adopted Egyptian habits, established their capital in Alexandria, and ruled until 30 BCE, when Egypt became a Roman province. Cleopatra VII was the final ruler of the dynasty.

KARNAK TEMPLE
One of the colossal statues at the Karnak temple, the largest in Egypt; during its construction, 30 Pharaohs came to power, including Hatshepsut, Seti I, Ramesses II, and Ramesses III.

The "War of the Impure," a "holy war" between rebellious followers of Set and the Theban clerics, worshippers of Amun, resulted in the division of the kingdom

The Nile

Natural resources and social organization go hand in hand throughout Egypt's history, as it was control of the River Nile's waters that led to the development of an empire whose backbone was formed by the river.

Sacred soil

The annual harvest was dependent on the flooding of the Nile in July and August each year. The "black soil" was the layer of dark mud left by the receding waters, fertilizing the land for agriculture. The rising of the river was just as important as the dykes, wells, and ponds that they built to control water resources that served to irrigate the fields in the succeeding months.

Fellahin. The term used to refer to peasants, fundamental to the Egyptian economy, whose prosperity depended on agriculture.

MAIN CROPS

- **Cereals:** wheat (to make bread); barley (to produce beer).
- **Pulses:** lentils and chickpeas.
- **Vegetables:** lettuce, garlic, celery, and onions.
- **Fruit:** dates in particular. The date palm was used to produce fibers.
- **Sesame plants:** from which oil was obtained.
- **Grapevines:** on the western delta. Wine was produced in the oases.
- **Linen:** for dresses, veils, and rope.
- **Papyrus plant:** from which the material for writing was made.

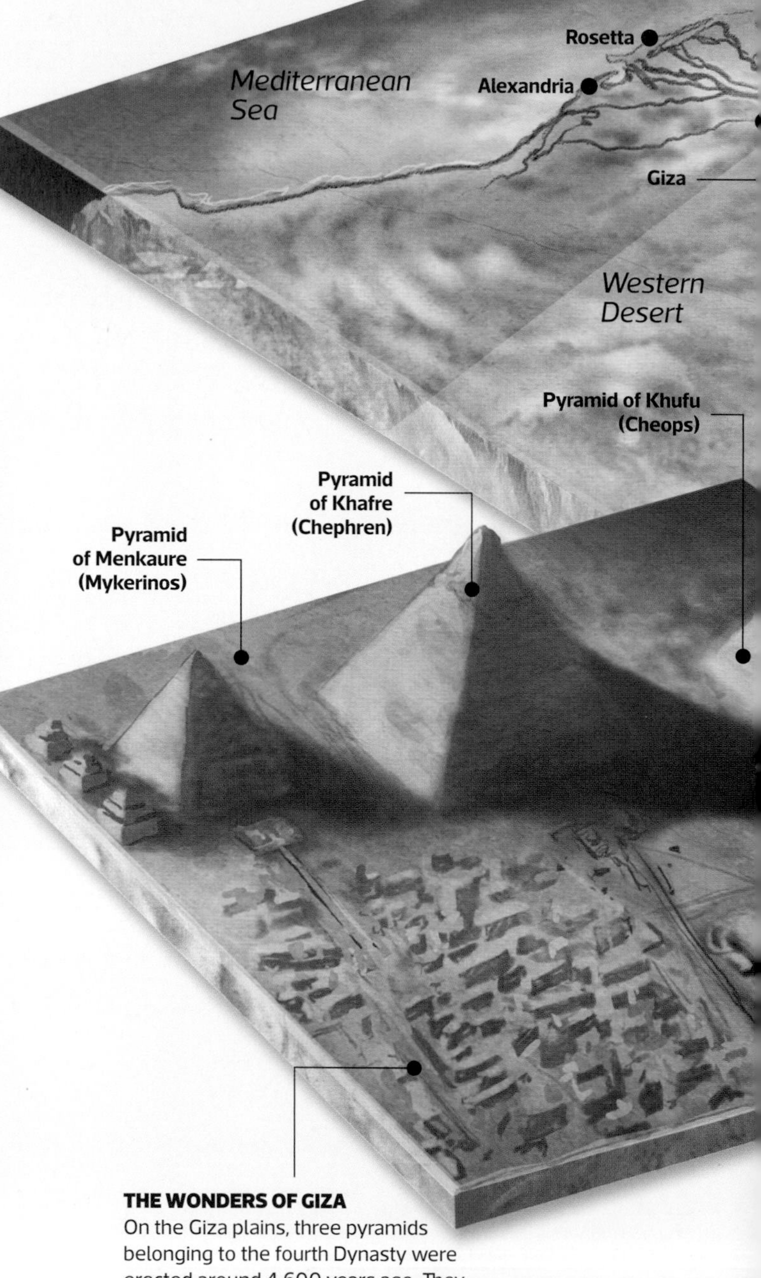

THE WONDERS OF GIZA
On the Giza plains, three pyramids belonging to the fourth Dynasty were erected around 4,600 years ago. They formed part of the great necropolis of Memphis, which included funerary complexes featuring temples, tombs, sphinxes, and pyramids.

Egypt at its most prosperous

The Nile Delta, whose lands were the most fertile, had the greatest concentration of wealth. Furthermore, Lower Egypt held the key that linked the empire to the Mediterranean, the main hub of trade traffic.

The two faces of Egypt

Upper and Lower Egypt were created as two different kingdoms. They were constantly in a state of confrontation until their unification around 3100 BCE. It is believed that Narmer, also known as Menes, was the king responsible for this event, and he founded the first of 31 Egyptian dynasties. Maintaining territorial unity was one of the most difficult tasks that the Pharaohs who succeeded him would face.

Nile Transportation

The River Nile was the link between the various Egyptian cities, from the Second Cataracts of Lower Nubia to the Mediterranean Sea. A whole host of vessels traveled the river, transporting people and goods from one side of the empire to the other.

LIFE ON THE BANKS
Flooding of the river was vital to the Egyptian economy. However, the huge floods affected the settlements on its banks.

Backbone

Over the centuries, Egyptian civilization gradually settled along the banks of the final 800 miles/1,300 km of the Nile. Farms dominated the landscape around its banks, and its waters were the primary means of communication. For daily tasks, small canoes were used; however, for trade or transporting passengers, strong sailboats were employed.

CANOES
Various different types made from reeds or papyrus, firmly tied together. They served as a means of exchange between traders and consumers.

POWER
In the canoes, passengers either sat to wield oars or remained on foot, pushing with long poles.

COMMERCIAL VESSELS
They traveled from port to port with soldiers and scribes on board. They sometimes measured over 130 ft/40 m in length, with a curved hull and sail.

Painted vessels

The importance of sailing to the Egyptian civilization can be seen in the appearance of vessels painted on the tombs of different Pharaohs.

Relief with a vessel on the Nile, featured on a private tomb in Saqqara.

PLOWS
The flooded, soft soil was plowed using draft animals.

SAIL
Square-shaped, made from papyrus fiber, and located on a central mast.

CANALS
Distributed the water as the Nile rose to fertilize the fields.

CROPLAND
Wheat and barley were produced in the irrigated fields and transported on both a small and a large scale.

FUNERARY MONUMENTS
The sophisticated pyramid construction required the transportation of stone over long distances.

HUNTING AND FISHING
Practised using canoes, with nets for fish and spears for aquatic birds.

LATTICE MASTS
Located at the bow and the stern to steer the vessel.

Woven reed huts

SCULLS
A pair of identical oars at the stern acted as rudders.

SHELL
Made from planks of cedarwood.

Barge for transporting heavy objects

Soldiers, animals, goods, gold, copper, and even stones and obelisks transported from granite quarries for funerary monuments were carried on large barges.

STRAKES
The rows of planks that covered the vessel's shell.

Buhen Fort

The Middle Kingdom was marked by the reunification of Egypt and the military incursions in the Nubian region, the purpose of which was to reestablish trade routes and strengthen borders. Buhen Fort was built as part of this offensive.

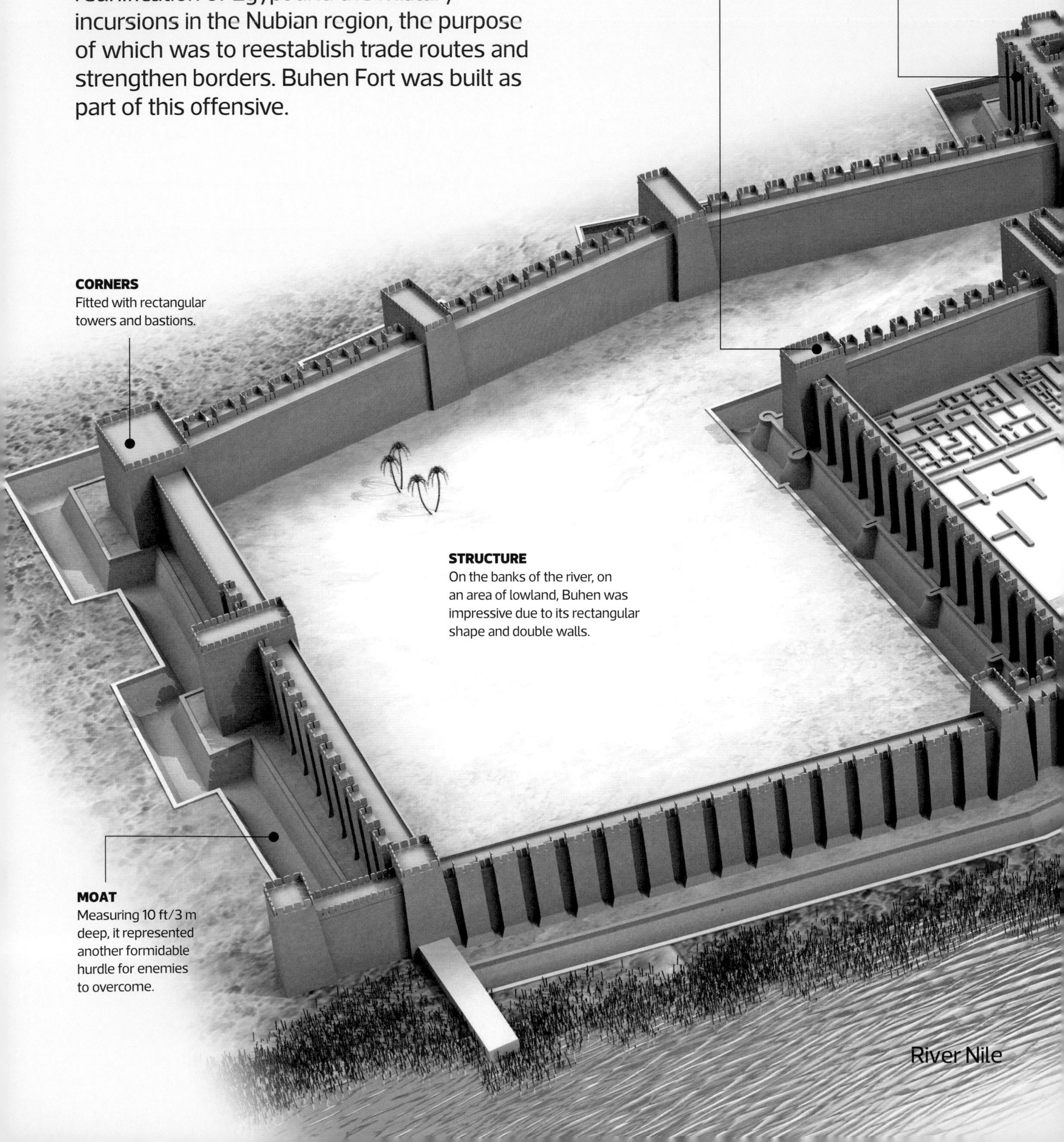

INTERIOR FORT
Protected by a wall measuring 490 x 560 ft/ 150 m x 170 m, it housed various outbuildings and a temple.

BARBICAN
Advanced fortifications, commonly used in medieval times, i.e. three millennia later.

CORNERS
Fitted with rectangular towers and bastions.

STRUCTURE
On the banks of the river, on an area of lowland, Buhen was impressive due to its rectangular shape and double walls.

MOAT
Measuring 10 ft/3 m deep, it represented another formidable hurdle for enemies to overcome.

River Nile

RESIDENTIAL AREAS
Residential spaces, the walls of which were found during nineteenth-century expeditions, can be seen in white.

TEMPLE
Dedicated to Horus, it was transported to the National Museum of Sudan in Khartoum in 1962, as a means of rescuing it from the artificial lake of the Aswan Dam.

The decorative paintings at this temple date back to the reign of Hatshepsut (1473–1458 BCE).

Most of the scenes at the temple show the king making offerings to the gods.

Residence of the fort's governor

HEIGHT
The walls measured around 36 ft/11 m high.

Docks

DOUBLE GATE
The monumental gates, located on the western face, opened toward the river. They were made of wood and featured a drawbridge.

OUTER WALL
Measuring 2,300 ft/700 m at its longest point, and 13 ft/4 m wide, numerous towers reinforced the fort's defense system.

FEATURES

- **Period:** between the twelfth and seventeenth Dynasties
- **Inhabitants:** it is believed 3,500 to 4,000 people lived here at its peak.
- **Discovery:** nineteenth century, during important rescue excavation work prior to the construction of the Aswan Dam between 1958 and 1970.
- **Location:**

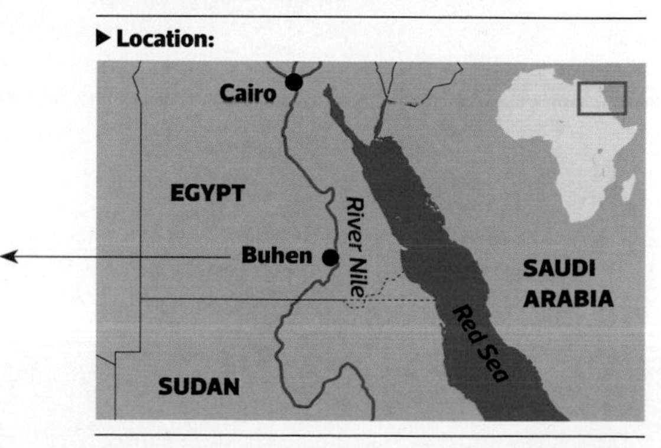

Pharaohs: Human and Divine

The monarchy was a fundamental pillar of Ancient Egyptian culture. The Pharaoh, who was both human and divine, was head of civil society, the military, and religion. He was responsible for the effective administration and defense of the country, in addition to maintaining the cosmic order.

OFFERING
Offering of light to Osiris in the darkness of the underworld.

The high priest

The divine nature of the Pharaoh made him the subject of worship and adoration. As a mediator between mankind and the gods, he was the high priest, the person responsible for the cult that helped communication with the gods. In this depiction, Ramesses III is at the transcendental moment, facing the judgment of the gods after his death on earth. The Pharaoh, as both human and divine, was assured of his acceptance into the afterlife.

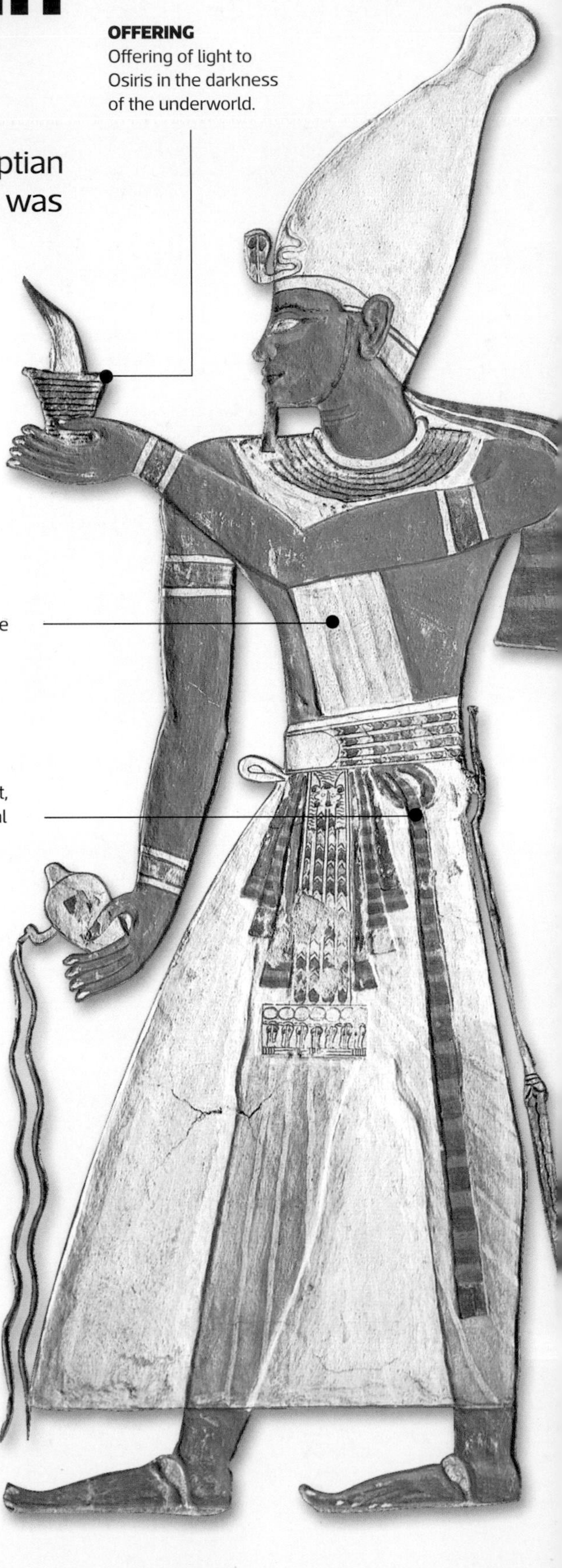

WRITING TABLET
Fastened to his waist, the tablet represents royal power through the ability to write.

CORD
Hanging from the waist, this represents physical strength and work.

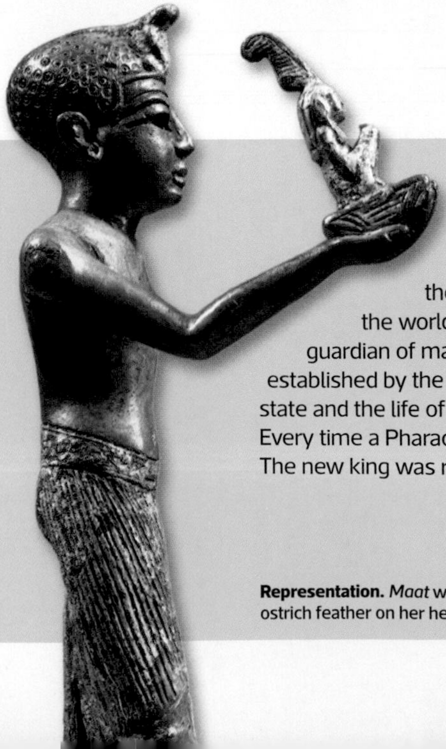

Maat

Maat is another fundamental concept of the Egyptian perception of the world. The Pharaoh is a transient guardian of maat, the state of harmony established by the gods on earth, on which the state and the life of mankind were dependent. Every time a Pharaoh died, maat disappeared. The new king was responsible for restoring it.

Representation. *Maat* was depicted as a goddess with an ostrich feather on her head.

The Triad of Menkaure

This sculpture depicts the three figures that represent the foundation of the political (Pharaoh), administrative (the goddess of the Cynopolis nome, or province), and religious (the goddess Hathor) systems. The triad is the expression of the authority that ruled Egypt; it went beyond earthly and human constraints, lasting for eternity.

THE GODDESS HATHOR
Goddess of the monarchy, deity of love and maternity, she is represented by the horns of a cow and a solar disk.

CROWN
Menkaure wears a white crown, the symbol of Upper Egypt.

ROYAL BEARD
Along with the crown, it symbolizes the power of the Pharaoh.

UNDER GUARD
The two female figures hold onto the arms of the Pharaoh using their hands, protecting and assisting him on his journey to eternal life.

FALCON
Symbolizes the attributes of the Pharaoh; as a bird of prey, it is associated with flying at heights and hunting on earth.

The Symbols of Power

The figure of the Pharaoh was unquestionable. However, his image had to transmit the values of divinity and, at the same time, political authority. Dynasty after dynasty, the symbolism of the Pharaoh's regalia remained almost unchanged.

Crowns

In Ancient Egypt, crowns served both political and religious functions; from the New Kingdom onward, they also served to establish the dynastic origin of the sovereign. In addition to the royal crown, there were also crowns used strictly for ceremonial or religious purposes.

DESHRET
This was the crown of Lower Egypt. It was red, with a conical base and a narrower part that jutted out to the back, with a spiral on the front that turned inward.

HEDJET
A long crown that stretched upward, toward a spherical top. It represented the monarchs of Upper Egypt and was the crown of Theban dynasties.

KHEPRESH
The blue crown that served a ceremonial purpose. The Pharaohs wore it during offerings to the gods.

PSCHENT
Comprising two long feathers, this was the crown of Amun. It also symbolizes the union of "Two Lands." During the New Kingdom, the women of the royal house and a number of priestesses wore this crown.

ATEF
Another ceremonial crown, used in religious contexts. It was related to Osiris and the god Heryshaf, with the head of a ram.

Symbols of authority

The authority of the supreme monarch was not just identified by his crown. Insignia, scepters, diadems, headdresses, and other lavish adornments were part of a wide range of items used by the Pharaoh and his family to maintain a respectful distance between them and their subjects.

NEMES
A fabric headdress that replaced the crown during the Pharaoh's daily activities. It was fastened to the Pharaoh's head using a diadem.

DIADEMS
Served to exhibit royal dignity, with or without the *nemes*, or royal headscarf. They were commonly used by the Pharaoh's children.

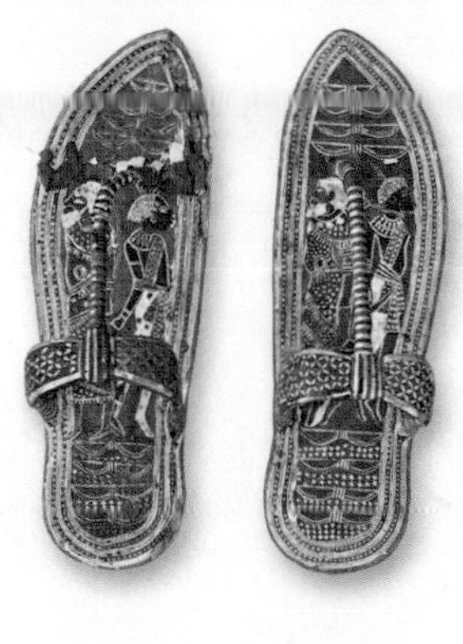

OSTENTATIOUS
Lavishness, as can be seen in these sandals belonging to Tutankhamun, was the common feature of royal attire.

URAEUS
The representation of Wadyet, the cobra goddess, protector of the Pharaohs, who were the only ones that could display it on their clothing.

SEJEM SCEPTER
Symbolized force and the magical energy of the king, his family, and the nobility. In this image, Queen Nefertari can be seen with the *sejem*.

RITUAL BEARD
This fake beard was worn by the Pharaoh on important occasions. It linked him to Osiris, the mythical founder of Egypt.

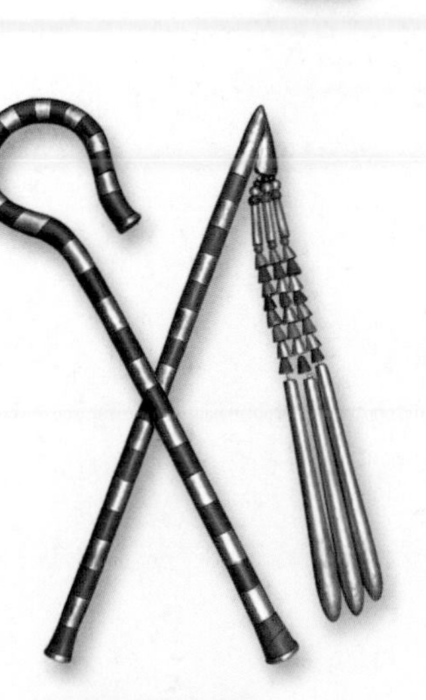

STAFF AND WHIP
The staff *heka* and the whip *nejej* are symbols associated with royalty. These pastoral instruments highlighted the Pharaoh's role as shepherd of his people.

State Administration

The Egyptian state was an absolute monarchy, over which the Pharaoh assumed all political, legal, and religious power. To exercise this power, a bureaucratic and administrative state was built around the sovereign that would gradually increase in complexity.

The role of the vizier

The vizier was similar in stature to a present-day prime minister. He was the chief official of the state and, as such, responsible for coordinating and controlling the entire Egyptian administration. Regional officials worked under his instruction.

FROM GENERAL TO PHARAOH
Horemheb, chief of the army during Tutankhamun's rule, was the last Pharaoh of the eighteenth Dynasty. After returning from his campaigns victorious, he married the sister of Nefertiti to legitimize his ascension.

WEAKNESS
In periods when the state was weak, the viziers, the heads of the army, and the nomarchs bolstered their power until they eventually occupied the role of Pharaoh.

Nomarchs

Egypt was divided into 42 nomes, or provinces. Each nomarch (provincial governor) acted on the orders of the vizier and was responsible for collecting taxes and recruiting peasants forced to work for the state. He was also responsible for maintaining the regional land registry.

MAGNIFICENT TOMBS
The impressive tombs belonging to some of the nomarchs are a testament to the power and independence that the people who occupied this role enjoyed during certain periods.

Privileged classes

The family of the monarch, the vizier and his relatives, local governors and army chiefs, together with those in the highest echelons of the clergy, formed part of the upper strata of the Egyptian hierarchical pyramid. Despite their privileged status compared to ordinary members of society, their life and work was dependent on the Pharaoh.

THE ROYAL FAMILY
Members of the royal family served as the Pharaoh's main advisers. Later, these positions would be occupied by people appointed by the Pharaoh himself.

THE OWNER OF EVERYTHING
In funerary stelae, the Pharaoh can be seen taking possession of goods owned by his subjects in life—he was the ultimate owner of everything.

HIERARCHIES
The hierarchical structure that divided Egyptian society was also preserved among the dead. In this wall painting dated to 1160 BCE, the deities that ruled the afterlife are larger than their subordinates, who also appear naked.

The architect vizier

Hemiunu was a vizier who answered to Khufu. A member of the royal family, the Pharaoh entrusted him with the construction of the pyramid at Giza. Archaeologists have found references to him as the supervisor of the works, holder of the royal seal, head of the army, and guide of the expeditions.

The Priestly Caste

In a civilization in which religion was so important, the priestly caste occupied one of the highest rungs in the upper echelons of power. Its members were responsible for managing the temples and dedicated their lives to worshipping the cult. Priests were greatly feared and respected by those who wished to be successful in crossing over to the afterlife.

TABLE OF OFFERINGS
As can be seen above, the table designed for offerings to the gods had different sections in the form of plates, on which different kinds of objects were placed. At the center, an inscription containing a number of hieroglyphs can be seen.

High priest

Meryre began his career under the Pharaoh Akhenaten as a temple steward and scribe, and became high priest of Aten. His titles included "Fan-bearer at the King's right hand" and "Sole companion." When the cult of Amun was restored in the reign of Tutankhamun, Meryre and his wife, Tener, disappeared and their splendid tomb was left unfinished.

Sculpture of Meryre and his wife, Tener.

PRIESTLY RULES
While they were responsible for rituals at a temple, priests were also obliged to abide by certain rules, such as eating little, abstaining from sexual relations, and performing their ablutions several times a day.

Egyptian temples

An Egyptian temple was designed as the house of god for a locality, or for the state in which it had been erected; it was not seen as a place at which believers should gather. Only priests were allowed to enter the sanctuary, where the god was incarnated in the corresponding statue. Their mission was to satisfy the ritual needs of the statue, with a whole host of ceremonies and offerings. They dressed the statue and fed it on a daily basis. They were also responsible for managing the temple's possessions.

KARNAK COMPLEX
Primarily dedicated to the cult of Amun, it was the most influential religious site during the New Kingdom.

A MAN OF POWER
The arrogant and solemn appearance of Ka-Aper in this statue would seem to suggest he held a high position in society. This contrasts with the simple nature of his clothing.

Scribe, military man, and priest

A scribe and lector priest during the fourth and fifth Dynasties (Old Kingdom), Ka-Aper fought in Egyptian military campaigns in Palestine, where he also served as a scribe in the Pharaoh's army. He was buried at Saqqara, along with a wooden statue depicting him and a bust that depicted his wife.

SHAVING
Given its polished surface, the statue of Ka-Aper appears to replicate the practice of shaving to which Egyptians subjected themselves on the grounds of hygiene.

CHIEFTAIN'S STAFF
This was an indisputable symbol of power. Without a doubt, the inspiration for this instrument came from the shepherd's crook that the Egyptians, like so many other civilizations, preserved from their origins as nomadic herdsmen.

Statue of Ka-Aper, chief lector priest (ca. 2500 BCE).

The journey to the afterlife

While the priestly caste governed all the laws and life cycles of animals and humans, their most important role was following the death of a person. This was perceived as an eventful journey that a person's spirit underwent while traveling to the afterlife. To successfully accomplish this task, it was essential that a person's body was purified by means of mummification.

A wall painting from the New Kingdom, depicting the funeral of Tutankhamun.

Sarcophagus of a priest of the cult of Amun.

The Scribes

Specialists in the art of writing, they formed part of the public administration and were indispensable to the Pharaoh. Among other responsibilities, they were tasked with bookkeeping and collecting state administration tax, documenting social events, and classifying and copying all types of information.

WRITING KALAMOS
A hollowed-out reed cane, cut on a slant and dipped in ink.

APPROPRIATE POSTURE
They sat with their legs crossed and placed the papyrus on top of them. Some chose to work in a crouched position.

Commanding respect

In a practically illiterate society, the scribes' abilities were admired and rewarded. Among the privileges they enjoyed was an exemption from paying tax. In turn, the scribes were incredibly strict and most of the population feared them. Numerous Pharaohs were unable to read or write; as a result, professionals in the art of writing had to be completely trustworthy. Three types of script were used: hieroglyphs, demotic, and hieratic.

Statue of a scribe sitting with a papyrus on his lap and holding a quill. Found in Saqqara, from the fifth Dynasty.

Learning the words

Becoming a scribe involved undergoing expensive preparation over many years. Therefore, only the sons of nobility or corresponding sections of society could afford such training. At school, located in temples or palaces, very harsh teaching methods were used.

FROM THE NILE TO THE SCRIBES
Papyrus was obtained from the plant of the same name that grew on the banks of the River Nile.

CUSTODIANS
Scribes were responsible for monitoring the country's intellectual production.

INKS
Red- or black-colored inks made from natural extracts were distributed on a palette.

Work and Slavery

Priests, the army, officials, and large-scale traders enjoyed a solid social status in Egyptian society. In contrast, small-scale traders, artisans, and peasants led a miserable life. Slaves occupied the lowest rung on a significantly divided social ladder.

Domestic service. A slave assists a wealthy lady to put on her necklaces.

The Egyptian economy

The significant efforts involved in constructing the Egyptian pyramids and temples are a reflection of the wealth of a country with an economy that was centralized and controlled by the state. This entity was responsible for providing workers and their families with a salary and subsistence in exchange for their work, which was strictly controlled by the officials.

FREE WORKERS
The discovery of the burial site of workers who contributed to the construction of Khufu's pyramid has helped to confirm that workers were free individuals, and not slaves.

Slavery

Slavery was a part of Egyptian society. However, it was less common than in Greece and Rome, and involved different characteristics. In general, slaves were destined for domestic service and the production of luxury goods. They enjoyed some rights, although they could be bought and sold. Foreigners were treated worst; most were prisoners of war who were sent to work in the mines.

CONSTRUCTION WORKERS
Fixed-term laborers and temporary workers, who generally came from the agricultural sector, worked on large-scale buildings when they were unable to work the land.

By sectors

OFFICIALS
The state retained a strong body of officials on its payroll. They enjoyed a good standard of living, as they had access to culture and education. Scribes, doctors, mathematicians, architects, and other professionals formed part of the higher levels of Egyptian officialdom.

ARTISANS
Artisans who worked in the workshops of royal palaces and temples enjoyed a better standard of living than those who worked in the market or ran roadside stalls. Such vendors belonged to the lower rungs of society.

FARMERS
Agriculture was the cornerstone of the Egyptian economy. Agricultural activities included raising animals, especially cattle, to produce meat, milk, and leather. Most of the fruits of their labor were handed over to the Pharaoh.

Military Power

Egypt, as a vast, rich empire, required several armies capable of waging war on different fronts simultaneously. This military power also guaranteed the unity of Egypt, where the rigid, centralist system often saw local forces participating in uprisings.

A knife and its gold-plated sheath, dated to ca. 1347–1337 BCE. They were found in the tomb of Tutankhamun.

Ramesses II during the Battle of Kadesh (1275 BCE).

NUBIAN ARCHERS
The *Medjay* archers (Nubian, from the Medjay tribe) were used as mercenaries to preserve domestic order. Other Nubian archers were responsible for guarding the borders.

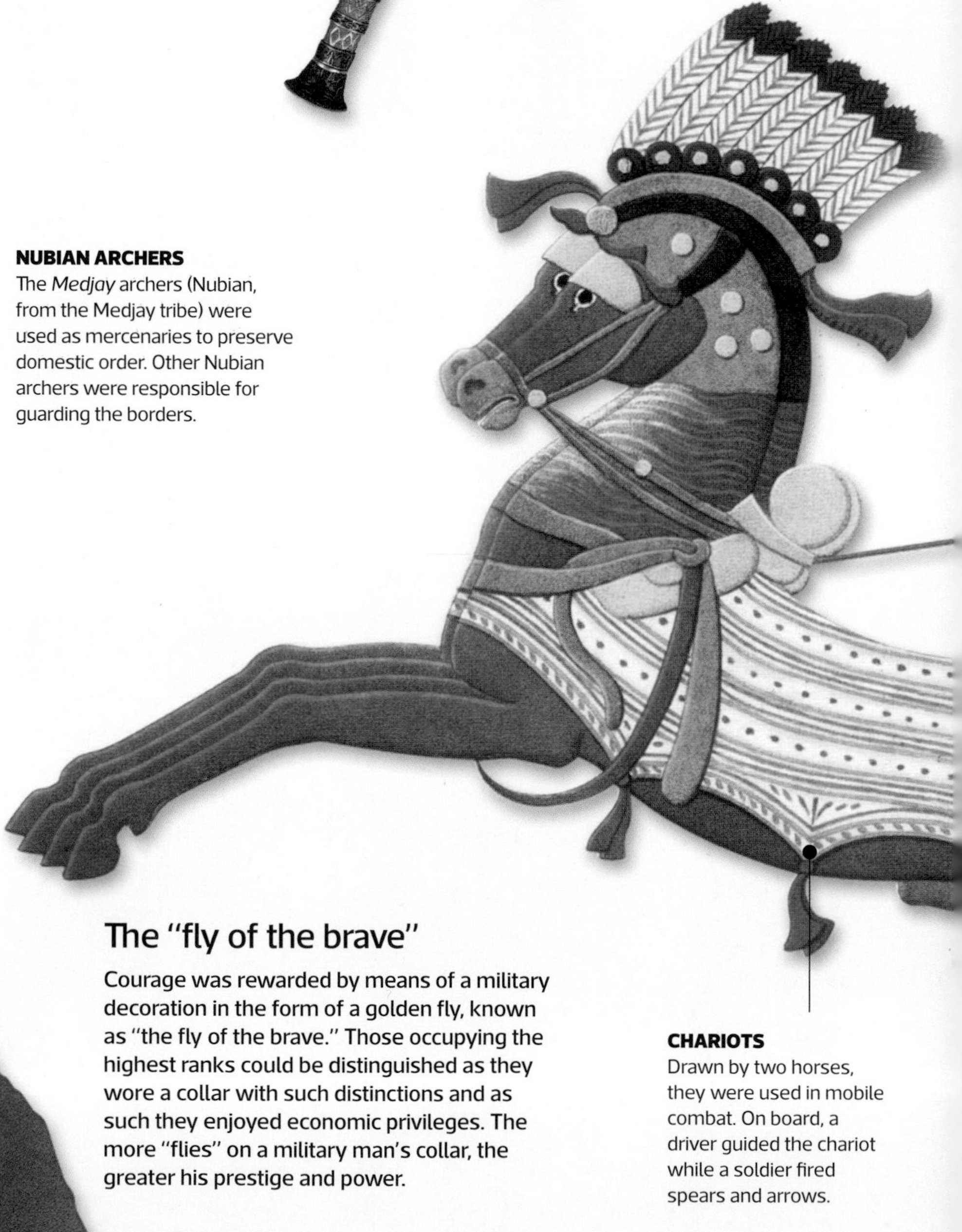

The "fly of the brave"

Courage was rewarded by means of a military decoration in the form of a golden fly, known as "the fly of the brave." Those occupying the highest ranks could be distinguished as they wore a collar with such distinctions and as such they enjoyed economic privileges. The more "flies" on a military man's collar, the greater his prestige and power.

CHARIOTS
Drawn by two horses, they were used in mobile combat. On board, a driver guided the chariot while a soldier fired spears and arrows.

THE BOW AND ARROW
The shaft of the arrows was made from reeds, at one end, a metal point was inserted and fastened, using resin and linen string. At the opposite end of the arrow, a slit was carved out to insert the bowstring.

Horus, the marksman

Horus was given this title as he was considered the inventor of the bow. Archers occupied a special position within the Egyptian army, performing decisive tasks in offensives and during sieges on enemy cities.

Carved onto the tomb of Ramesses II, he is depicted inflicting death on his enemies (1257 BCE).

ARCHERY
Considered a religious activity. During the festival of Heb Sed, the Pharaoh shot arrows in the direction of the four cardinal points, to show that his power defended the entire empire.

SUPREME CHIEF
The Pharaoh was the supreme chief of the army, although he relied on intermediary generals and officials. Officials used a long baton that distinguished them from the other soldiers.

Great expansion

The New Kingdom coincided with a period of significant growth. Thutmose I punished the peoples of Nubia and swept Syria and Palestine aside until he reached Mesopotamia. Thutmose III maintained persistent campaigns to extend the conquests that, in addition to providing resources, resulted in many enemies being taken prisoner.

Obelisk dedicated to Thutmose III.

Egyptian Soldiers

The Egyptian army did not turn professional until the Middle Kingdom. It experienced significant growth during the Second Intermediate period, following the invention of chariots of war, and reached the height of its professionalization at the beginning of the New Kingdom, a period of military expansion.

Mercenary troops

The Pharaoh's army featured mercenaries, most of whom were Nubians (below), famed for their archery skills, and Libyans. The Nubians used special arrows with a flint tip. The *sherden*, one of the so-called "Sea Peoples," served as the Pharaoh's personal bodyguard.

The army of the New Kingdom

The Egyptian armies of the New Kingdom consisted of huge units of spearmen supported by a similar number of archers, in addition to chariots of war. The way the Egyptian infantry fought was quite simple: archers launched a barrage of arrows on their enemies, after which the spearmen charged. As well as wielding their spears, they used hand weapons, such as axes, knives, and maces, for close combat.

WAR EMBLEMS
The standard-bearers (*tay-seryt*) led the troops, bearing the insignia of each company. The most common standard was in the form of a fan.

Their role was to keep the troops organized.

They tended to feature the image of a sacred animal.

Other weapons

Egyptian soldiers were trained to use a whole host of weapons, although specialization in one particular weapon was essential for creating companies.

BOWS
The compound bow, inherited from the Hyksos, was highly regarded during the New Kingdom.

DAGGER
With a stone or bronze blade, it was the hand weapon of choice among archers.

Marching, instruction, and combat

Soldiers left their homes behind forever, as it was more than likely they would not live to see them again. A harsh training regime awaited them, involving desert marches and close combat.

SPEAR
Ordinary troops used a wooden spear, similar in size to the height of its user.

SHIELD
Made from wood, the front was reinforced with leather and a bronze plate.

ARMOR
Made from several layers of reinforced linen, offering protection from arrows.

SWORD
The *khopesh* was made from bronze, and was sharpened on the convex edge.

ARROWS
The tips of the arrows were made from flint or bronze.

PROTECTION
Made from hardened linen, it was rigid and as resistant as leather.

AX
One of the most commonly used models was the epsilon ax, with a bronze blade.

MARCHING
During campaigns, troops marched an average 12 miles/ 19 km per day over desert terrain.

SPEARHEAD
During the New Kingdom, the tips were made from bronze or flint.

The Battle of Kadesh

With a view to recovering lands previously ruled by Thutmose III in Asia Minor, Ramesses II faced the Hittites at the Battle of Kadesh (1275 BCE), which did not result in a clear victory. It is remembered as the first battle involving chariots of war.

Details of the battle

The Battle of Kadesh was the first well-documented battle. The strength of the Pharaoh was highlighted in the *Poem of Pentaur*, carved into the walls of several temples. There are also other reliefs that recreate different events that occurred during the battle.

CHARIOT-ON-CHARIOT PLAN OF ATTACK
The fight between chariots required close proximity, to bombard the enemy's chariot, and distance, in order to reload the bow and resume the attack.

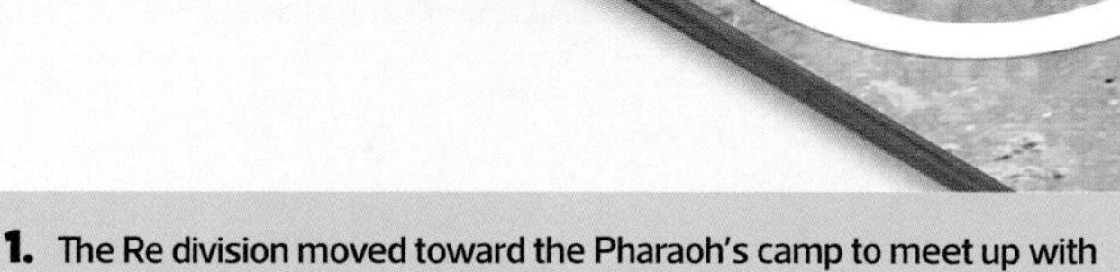

Risky maneuver

Ramesses II organized his troops into four divisions. The Amun division advanced in the belief that the Hittites were farther to the north. However, the Hittites, led by King Muwatalli, launched a surprise attack on the Pharaoh's camp. Thanks to a quick reaction and their chariots of war, the Egyptians were able to repel the Hittite attack.

BALANCE OF POWER

Egyptians	Hittites
▶ **Men:** 20,000	▶ **Men:** 40,000
▶ **Chariots of war:** 2,000	▶ **Chariots of war:** 3,500

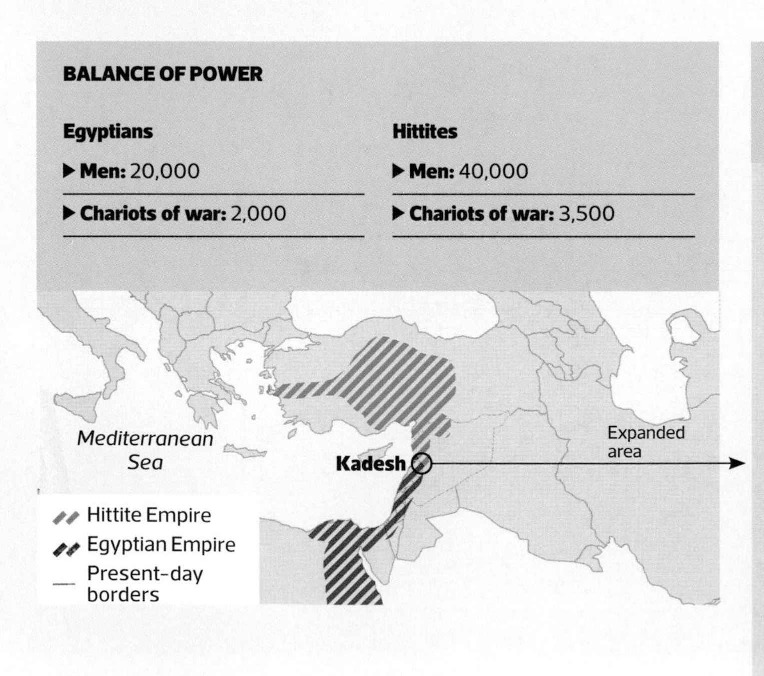

1. The Re division moved toward the Pharaoh's camp to meet up with the Amun division. However, the Hittite chariots of war charged one side of the division, taking them by surprise and causing them to flee.

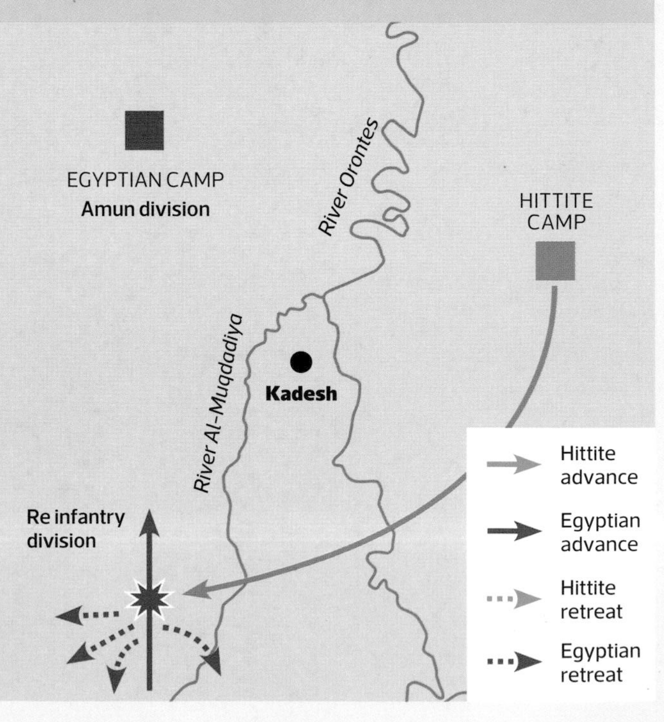

The outcome

Although the Egyptians were successful in driving away the Hittites, they were unable to gain control of Kadesh. Fifteen years later, both parties met again at the city to sign the first nonaggression pact known to humankind.

INFANTRY PLAN OF ATTACK
Chariots lunged at the infantry, dispersing its members. Finally, they doubled back on themselves and resumed the attack.

2. The Hittites reached the Pharaoh's camp and ransacked it. However, their heavy and slow chariots bunched together, making maneuvering impossible.

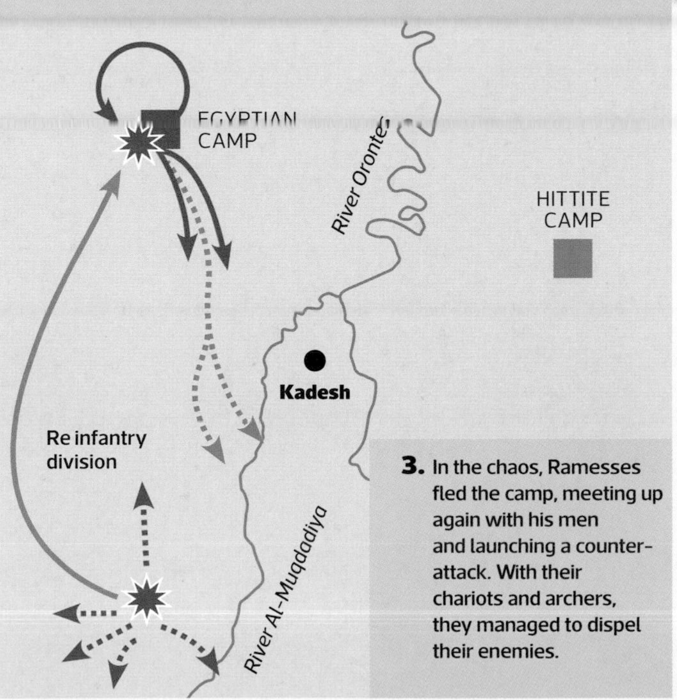

3. In the chaos, Ramesses fled the camp, meeting up again with his men and launching a counter-attack. With their chariots and archers, they managed to dispel their enemies.

4. Muwatalli sent in more troops, but the Amorite reinforcements serving the Pharaoh arrived from the north, neutralizing them.

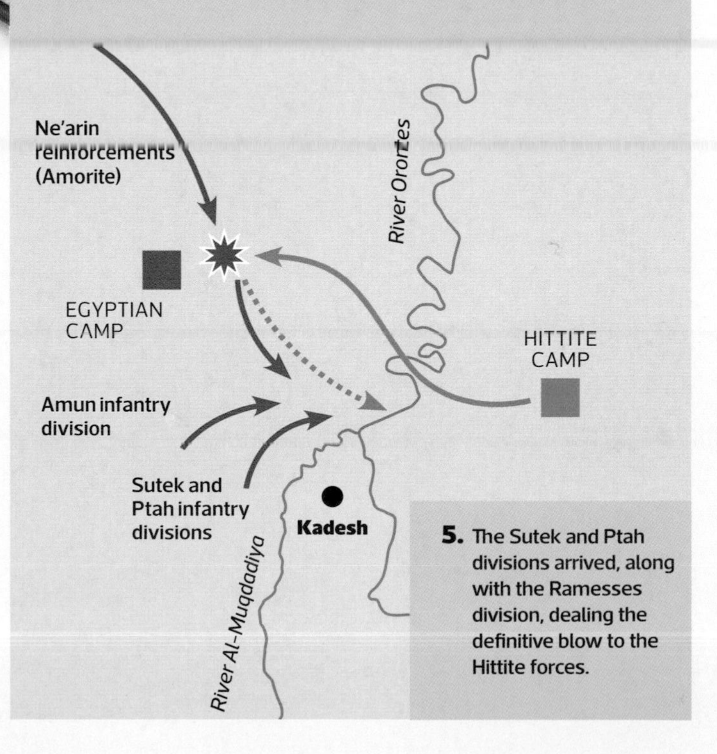

5. The Sutek and Ptah divisions arrived, along with the Ramesses division, dealing the definitive blow to the Hittite forces.

The Hittite Army, the Enemy of Egypt

As part of the Egyptian expansion into the northeast between the fifteenth and eighteenth centuries BCE, the Hittites were confronted. With its powerful army, this Indo-European people occupying the Anatolian Peninsula was considered a great power at the time.

HORSES
The Hittites used horses to draw their chariots, rather than donkeys or onagers as used by other peoples. Their mobility and power gave them a decisive advantage during combat.

Military innovations

The Hittites made several decisive innovations in terms of warfare. They substituted the use of bronze with iron to produce their weapons, helmets, and shields. Their chariots of war had two wheels with spokes, rather than the four solid wheels typical of the period, which made them faster and lighter.

GENERAL CHARACTERISTICS

Insignia. The Hittite people were the first to use a two-headed eagle as their national symbol.

- **High command:** King of Hatti
- **Structure:** infantry and chariots
- **Personnel in the field:** 40,000 or more
- **Chariots:** 3,000 or more

The last great king

Tudhaliya IV, who ruled between 1237 and 1209 BCE, was the last great Hittite king. Under the rule of his son, Suppiluliuma, the empire collapsed following invasions by the Sea Peoples.

SQUIRE
Three men traveled on each chariot of war: a charioteer, an archer, and a squire. The squire protected the warrior and equipped him with weapons.

WARRIOR
He was the only one to use plated armor and an iron helmet. He was capable of launching spears with a great degree of precision, and was also an expert archer.

QUIVER AND ARROWS
Although they transported quivers with bows and arrows on the chariot, the weapon of choice among Hittite warriors was the knife or ax.

CHARIOTEER
His role was vital, as the success of any attack depended on his driving expertise. The charioteer in the illustration sports the traditional haircut of Hittite warriors.

CHARIOTS
Structurally solid, but very fast compared with those used by their enemies. Their power broke enemy infantry lines.

Allies

The Hittite armies' lines featured elite troops, such as the Royal Guard or Meshedi; and city-state lieges, such as Mitanni, Ugarit, or Luqqa.

Royal Guard

Mitanni

Ugarit

Luqqa

Chariots of War

The introduction of chariots of war on the battlefield changed military techniques. They were first used by Mesopotamian warriors around the second millennium BCE, but the Egyptians improved the design and gained stability, strength, and safety.

Chariots on the battlefield

Given their agility and speed, chariots were used to take up positions on the battlefield and facilitate the use of bows and arrows and spears. They served as infantry reinforcement units, and were the foundation of the Egyptian army. They were also very useful when chasing a fleeing enemy.

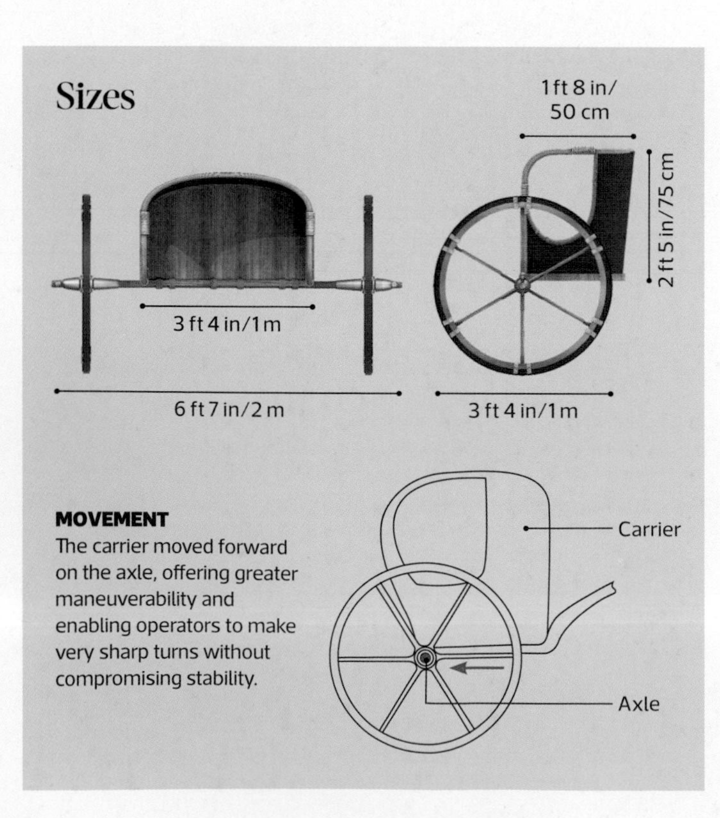

Technical development

The Egyptians perfected the chariots employed by Hyksos invaders, converting them into versatile mobile platforms for shooting arrows and throwing spears.

- **Types of wheel:** spoked, made from wood and iron
- **Number of wheels:** 2
- **Driven by:** 2 horses
- **Crew:** 2

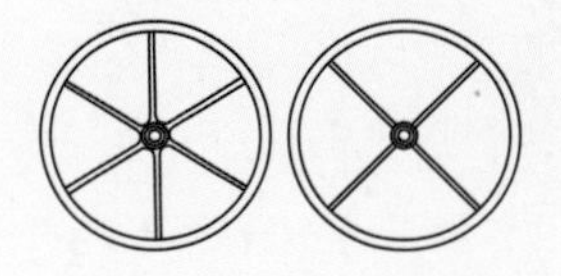

QUIVER
Made from wood and leather, it was firmly attached to the chariot's carrier.

SHAFT
Made from a single piece of wood.

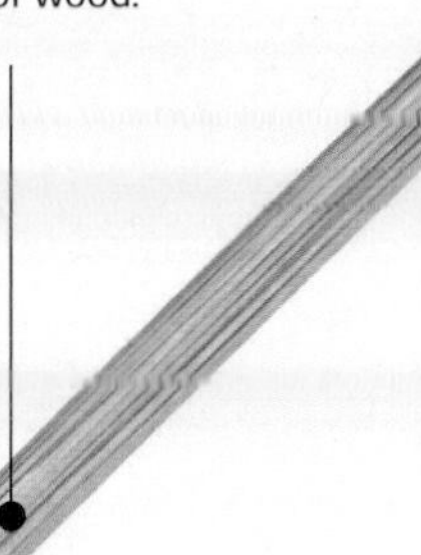

THE CARRIER
Comprising curved wooden armor and reinforced using leather and copper bolts. Fitted to the shaft and central part of the axle with straps.

SECURITY
The wheels were fastened using a locking bolt that passed through the axle.

Other chariots of war

THE SUMERIANS
Credited with having invented the chariot of war, in addition to using them for transportation purposes. Their use became more general in wars fought by the city-states of the region.

- **Types of wheel:** solid, made from wood
- **Number of wheels:** 4
- **Driven by:** 4 wild donkeys
- **Crew:** 2

THE HITTITES
Present-day Anatolia was the heart of this civilization between the eighteenth and twelfth centuries BCE. The Hittites perfected the light chariot and made it their main weapon.

- **Types of wheel:** spoked, made from wood and iron
- **Number of wheels:** 2
- **Driven by:** 2 horses
- **Crew:** 3

THE ASSYRIANS
Around the eighth century BCE, the Assyrians used robust chariots of war, with large wheels and shields, most likely used as a battering force.

- **Types of wheel:** spoked, made from wood and iron
- **Number of wheels:** 2
- **Driven by:** 2, 3, or 4 horses
- **Crew:** 4

The Battle of the Delta

The first naval battle since records began occurred in northern Egypt around 1176 BCE on the Nile Delta. The Egyptians faced the "Sea Peoples," so-called as it is not known where they came from. They were responsible for destroying the Hittite Empire and other kingdoms.

1. It is believed that the fleet of the Sea Peoples was large, and that its sailors were experienced. Its soldiers used large spears and round shields with a central handle.

2. Aware that their enemy was seeking to invade and occupy land, the Egyptians ambushed their opponents, enticing them toward the branches and canals that formed the delta of the River Nile.

Naval battle

The Battle of the Delta was a bloody and cruel contest in which the fleet of Ramesses III faced the Sea Peoples, although the infantry of the Egyptian navy would also take part. Despite the Egyptians' clear triumph, their economy was weakened; this situation would last for the remainder of the New Kingdom.

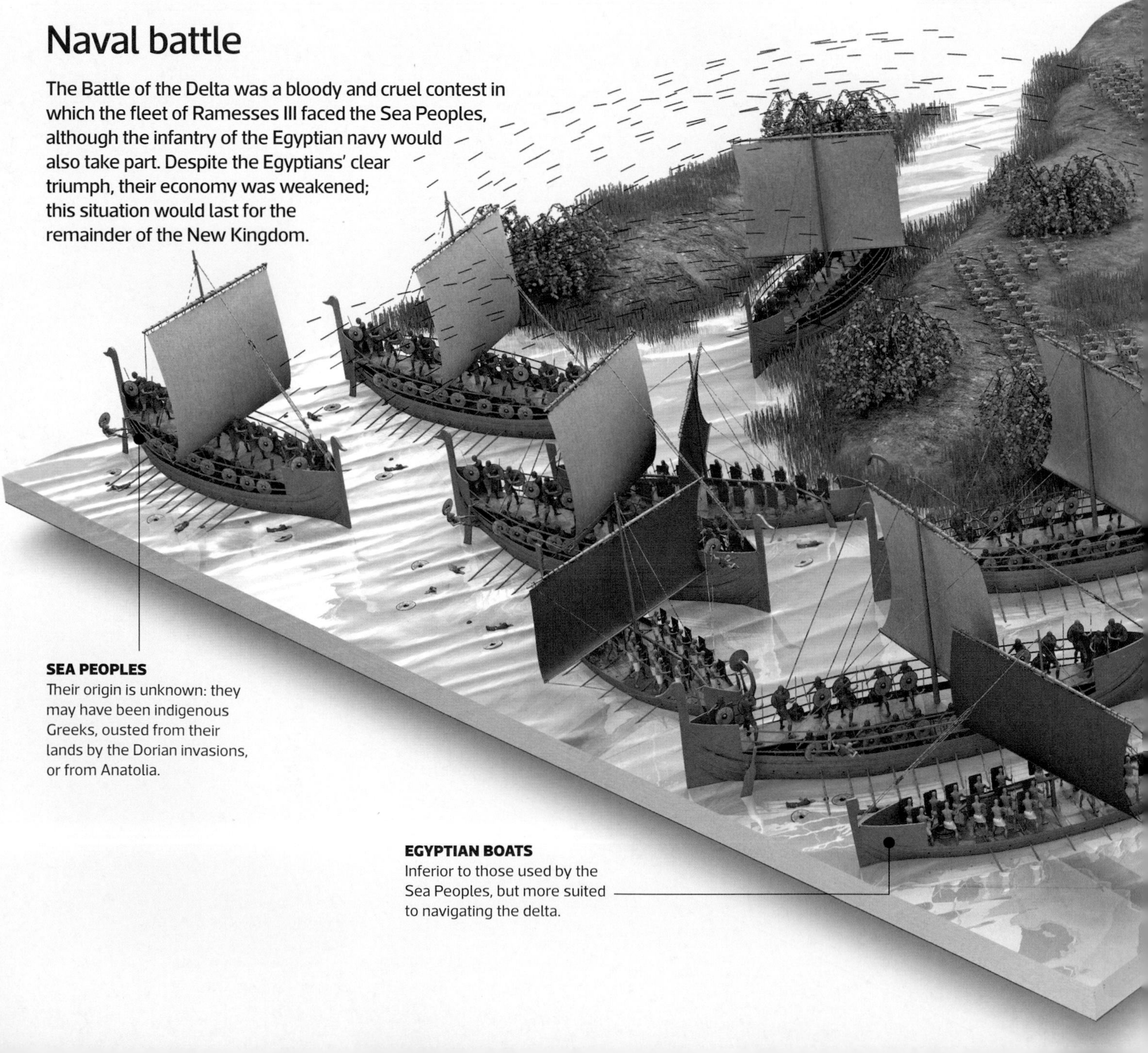

SEA PEOPLES
Their origin is unknown: they may have been indigenous Greeks, ousted from their lands by the Dorian invasions, or from Anatolia.

EGYPTIAN BOATS
Inferior to those used by the Sea Peoples, but more suited to navigating the delta.

3. Once in the delta, the invaders were welcomed by a barrage of arrows launched from the Egyptian ships and from dry land. The Egyptian infantry also planned for close combat with the enemy.

CANAL NETWORK
Crucial to the Egyptians' tactics. In this intricate network, the mobility of the enemy's fleet was restricted; this made an attack much more feasible.

INFANTRY
Records suggest that "the best of the Egyptian infantry" was selected for the battle.

HANDHELD WEAPONRY
It is known that spears and shields played a fundamental role in close combat, and that little mercy was shown to the losers.

BARRAGE OF ARROWS
One tactic was to entice enemies to the shore, to subject them to an intense barrage of projectiles: spears, javelins, and arrows, in particular.

Unique document

The great stele at the funerary temple of Ramesses III, in Medinet Habu, is the only document known with images of the Battle of the Delta. Otherwise, only various rare mentions of it are made in documents from the time.

War Boats

With a view to definitively defeating the Sea Peoples, who regularly besieged the Egyptians, Ramesses III decided to construct the war fleet that Egypt had lacked until that point in time. The "bireme" would prove to be a blueprint for future galleys.

TECHNICAL DATA

- **Origin:** Egypt
- **Displacement:** 10 tons
- **Crew:** 38 (a dozen of whom were soldiers)
- **Rowers:** 24
- **Notable involvement:** Battle of the Delta (1176 BCE)
- **Fleet:** Ramesses III

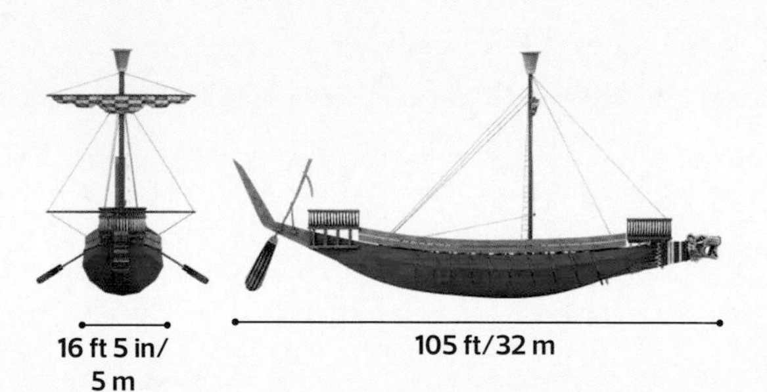

CENTRAL COVER
The archers were placed here, ready to fire. On both sides of the cover, seats were set out for the rowers.

BATTERING RAM
Made from wood and reinforced with bronze. Given its location at the bow, it has not been possible to establish with any certainty if it was used to charge at enemy vessels.

CROW'S NEST
This type of boat was the first known to feature a crow's nest. Made from woven papyrus, it topped the ship's mast.

FORECASTLE AND RUDDER
The boat features two wooden forecastles, where most of the soldiers congregated. A powerful oar to the aft served as a rudder.

SHELTERED ROWERS
A high, strong edge made from wood protected the rowers from enemy arrows and spears.

STRUCTURE
This type of boat had no keel or frame. It was made from blocks of acacia wood from the Nile area, bound together as if they were bricks.

Other boats

Later, the Egyptians were happy to copy designs from peoples with a greater naval experience, such as the Greeks, Phoenicians, and Romans.

QUINQUEREME

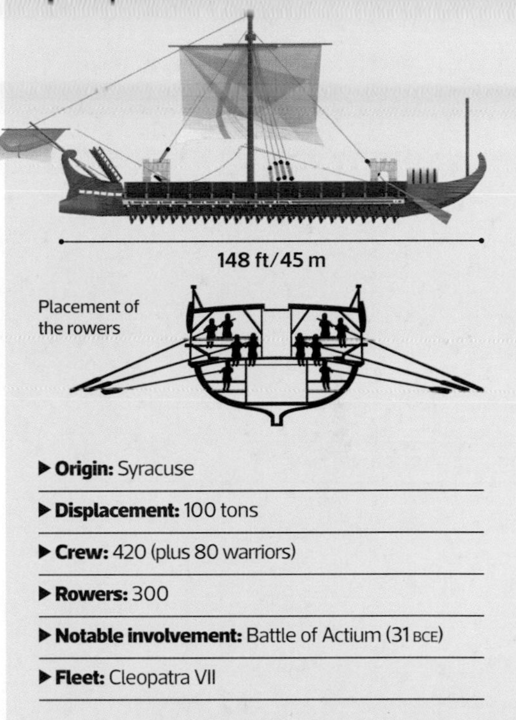

- **Origin:** Syracuse
- **Displacement:** 100 tons
- **Crew:** 420 (plus 80 warriors)
- **Rowers:** 300
- **Notable involvement:** Battle of Actium (31 BCE)
- **Fleet:** Cleopatra VII

TRIHEMIOLIA

- **Origin:** Rhodes
- **Displacement:** 40 tons
- **Crew:** 144
- **Rowers:** 120
- **Notable involvement:** Battle of Cos (261 BCE)
- **Fleet:** Ptolemy II

3 CHAPTER

SOCIETY AND DAILY LIFE

LIFE ON THE NILE

Thanks to archaeological findings, today we know that the Ancient Egyptians lived in spacious houses with functional furniture, took excellent care of their physical appearance, and were in constant contact with other peoples.

As researchers progress in their studies and discoveries, the Ancient Egyptian civilization becomes increasingly more fascinating. The majestic nature of its pyramids, temples, and sculptures captivates scientists and visitors alike, although it represents only the most obvious and visible aspect of a lifestyle that was distinguishable from other cultures by its unique features. The urban revolution, immediately prior to 3000 BCE, was the beginning of a transformation from large villages to towns—the chrysalis of the civilization's identity. Egyptian civilization progressively began to express itself through new forms of social, economic, political, and religious organization; this would encompass the old "nomes," a type of fiefdom that comprised a central village responsible for other peripheral settlements.

The Nile represented the backbone of this civilization, allowing communication between Lower Egypt, the heart of which was the productive delta, and Upper Egypt, which, being more distant from the trade routes of the Mediterranean, was poorer and less developed. Maintaining the unity of the empire was the main concern of the Pharaohs and was a problem that would run throughout the civilization's history. The Nile was initially crisscrossed by papyrus boats, although these soon gave way to wooden boats built in the shipyards on the river's banks. As can be seen in the frescoes that adorn the temples, larger and stronger boats transported huge granite pillars and obelisks from the Aswan quarries to destinations hundreds of miles away. Smaller trade vessels were used to transport grain and, furthermore, during conflicts they were used to transport military personnel, although the Ancient Egyptians did

DAILY LIFE A relief from a tomb from the eighteenth Dynasty (1550–1295 BCE), with scenes depicting daily life in a dwelling.

BOATS
The Nile was the backbone of the empire and the main channel of communication within the country and up to the Mediterranean Sea.

not develop the idea of a war fleet. Official vessels were designated for government personnel and even members of the royal family. As is the case in modern times, the launch of a new ship was cause for a special celebration, with representatives present from the highest levels of the administration.

Cool and spacious living quarters

As would be expected, different social spheres led different lifestyles, although there was a certain level of uniformity in many aspects. Houses in Ancient Egypt were built using sunbaked bricks made from mud collected from the Nile. This material was collected in drums or leather tubs, and transported to the construction site. Once there, workers added straw and pebbles to reinforce it, pouring it into wooden molds to give it shape. Once formed, the bricks were removed from the molds and left to dry in the sun. When construction on the house was complete, the walls were lined with plaster.

The temperature inside the dwelling remained cool, as the windows were placed high up the wall and were small in size, in order to prevent direct sunlight from entering. Most houses were rectangular in shape and had a floor area of between 1,100 and 1,350 sq ft/100 and 125 m^2. The walls of single-story houses were around 16 in/40 cm thick, but more than 5 ft/1.5 m thick on two-story constructions. Door frames were generally built from stone, even those installed in the poorest houses. Simple doors and double leaf doors were made from wood and could be locked from the inside. Keys from around 1500 BCE have been found that, given their simple appearance, must have corresponded to a very rudimentary lock mechanism.

ROYAL THRONE
Throne made from wood that belonged to Queen Hetepheres. She rested her arms on a lotus-flower motif.

In the dry and desert regions of Lower Egypt, water use was subject to strict control due to its scarcity and was probably a regular cause of serious conflict

Copper pipes that were used to drain water have been found, although only at certain temples; to date, they have not been found in standard housing. Most toilet seats were made from limestone; alternatively, a stool featuring a hole was used. Waste water was sent to special wells, the river, or the street. For other uses, water dripped into a barrel, which was later emptied by hand, or drained slowly into the soil through a hole at the bottom of the barrel. In the dry and desert regions of Lower Egypt, water use was subject to strict control due to its scarcity. Water control was probably a regular cause of serious conflict and confrontation.

The houses of more wealthy Egyptians were also made from adobe bricks and painted white, featuring one or two storage rooms for housing work tools, domestic utensils, and foodstuffs. The rooms and private spaces were behind a space that served as an entrance hall. A flat roof, laid over beams and accessed by a ladder, served as a terrace. Various types of shutters, covered by reed blinds, protected the inside of the house against sunlight. The areas located farther into the house were lit by a series of perfectly fitted skylights to prevent rainwater from entering. The upper classes tended to live in larger houses, with a garden featuring trees that offered shade and a pool to freshen the environment. They also featured a central inner courtyard, which served as another room in the house.

Trees and gardens

An inscription on an Egyptian tomb, dated to 1400 BCE, states: "May I walk every day unceasing on the banks of my water, may my soul rest on the branches of the trees I have planted, may I refresh myself in the shadow of my sycamore." In reality, growing plants and trees was common in Ancient Egypt, whether in private gardens or public spaces. Images of

SERVITUDE
Wealthy families owned slaves who performed domestic chores. This painted stone statuette, found in Saqqara, depicts a servant with a jar.

Furniture was often simple but perfectly designed; examples display not only functional comfort, but a certain level of taste and workmanship

AMULETS
The use of amulets was commonplace among Egyptians. This one was designed to protect women during pregnancy and birth.

these gardens are present on numerous tombs, featuring rectangular pools, often full of fish, and vines planted in straight lines. Trees and bushes were much sought after given the shelter they provided, especially during the hot seasons, and for their delicious fruits. This was especially the case for date palms and pomegranate and walnut trees and, to a lesser extent, willows, acacias, and other decorative species. On some frescoes, different types of flowers can be seen: such as daisies, mandrakes, roses, irises, myrtle, jasmine, narcissus, ivy, henna, and laurel.

Comfortable furniture

In general, the walls of houses were decorated with bright, colorful drawings of figures and geometric shapes. Simpler houses scarcely contained furniture. Long benches built into the walls served as a seat, table, and bed, simultaneously. In houses featuring furniture, it was often simple but perfectly designed; examples display not only functional comfort, but a certain level of taste and workmanship. Chairs with arms and backrests, and wooden stools, were used in abundance. In some unearthed frescoes, the existence of sofas with cushions, most probably stuffed with mallard feathers, can be seen. Wool, the most commonly used fabric along with linen, served in the manufacture of carpets and other upholstery in the home. It was rarely used for clothing purposes, as it was of animal origin and thus considered impure. Beds were often made from wicker or wood. On warm evenings, fortunate Egyptians would probably not use a pillow, but they would place their head on a shaped wooden headrest designed specifically for this end. Chests and drawers were used, most designed to store personal objects and utensils, in addition to tables that were often decorated with inlaid art. In a space designated for cooking and food, each family kneaded their own bread and produced their own beer. Numerous utensils used in these household chores have been preserved to this day.

GRINDING
Statuette of a woman grinding grain with a stone, dated to the fifth Dynasty, during the Old Kingdom.

Family

The home served as a meeting place for the entire family; here, the youngest members would play while the older members rested during their free time. Furthermore, as illustrated on many frescoes, Ancient Egyptians often held parties at their homes, which were attended by men, women, and children.

Women were responsible for keeping the house and raising children, while their husbands worked outside the home. Marriage was a contract related to cohabitation as a couple, but it did not involve a sacred bond. Even though a spouse was suggested by her father, a future wife was able to make the final decision before deciding to marry. Likewise, if marriage did not work out, regardless of whether the couple was unable to have children, or whether one of the partners was unfaithful, divorce was an option. Due to the high level of infant mortality, couples tended to have many children. Monogamy was the most common option, as only the Pharaoh and some members of the court were capable of maintaining more than one wife at a time.

COSMETICS
Alabaster bottles dated to between 1500 and 1300 BCE that contained ointments used to spread on the skin.

Bustling cities

Although the Egyptian lifestyle was unique, habits and customs from other cultures also had an influence; this was particularly true of cultures with which the empire maintained a strong trade relationship. Constant exchanges were made with Babylonia and Assyria; this was also the case with the Phoenicians, Hittites, Israelites, and other Achaean and Aegean kingdoms or city-states based in Asia Minor or the Peloponnese. It is known that in around 1450 BCE, the Pharaoh received a number of Cretan and Mycenaean delegates, who brought important gifts with them. This also occurred when the Mycenaean Greeks took control of Cyprus and the western coast of Asia Minor. Various Hellenic sources spoke of the Greeks' surprise when they visited Egypt. They described huge cities with

CERAMICS
Zoomorphic jar in the shape of a fish, dated to between 4000 and 3200 BCE.

FEMALE SENSUALITY
Female torso dated to between 1567 and 1320 BCE. Semitransparent fabrics were often used in women's clothing to highlight the curves of their bodies.

busy streets and brightly decorated temples, as well as great festivals and ceremonies celebrated with great pomp. Egypt was the richest and most powerful country in the ancient world. Some of the gold from the mines in the eastern desert and Nubia was sent to faraway places, such as Mesopotamia, where it was bartered in exchange for manufactured products. Even though there were times when the Pharaohs controlled areas beyond the southernmost tip of the Nile, the products of equatorial Africa arrived in Ancient Egypt by means of trade with Nubian princes, who ruled the region south of the first cataract of the Nile.

Clothing, cosmetics, and games

The goddess associated with style and beauty was Hathor, worshipped as the epitome of elegance in various hymns and poems that praised her. Numerous objects have been found to this end, such as brushes, mirrors, and small cosmetic receptacles. Both Egyptian men and women used a variety of oils to cover their face and paint their eyelids. Cosmetic ointments were made using minerals crushed on fine slate palettes. The resulting powder was mixed with water or diluted oil and preserved in tubes. The mixture was extracted using thin sticks, which were also used in its application. Using malachite, a copper mineral, the Egyptians obtained a green paint that symbolized fertility. Galena, a leadlike mineral, produced a black paint, and was used to emphasize an Egyptian's eyes. For their cheeks and lips, Egyptians used red ocher, an iron oxide that was common throughout the empire.

As can be seen in excavated paintings, courtesans wore flower-shaped ribbons on top of their wigs, which were most probably held in place by perfumed beeswax, while courtiers often viewed themselves in finely

GAMES
A board featuring the popular game of *senet*, made from ebony and ivory, found in the tomb of Tutankhamun.

polished copper mirrors, which were also recreated in some funerary representations. Given the weather, the same style of light and cool linen clothing was maintained throughout history in Ancient Egypt. Free Egyptians wore a knee-length skirt. Slaves wore a loincloth or went completely naked. It was common for children to wear no clothing. Women wore a close-fitting dress that covered them from chest to ankles. This tight dress was still worn during the New Kingdom, although it was worn as underwear, when it became common to wear a pleated tunic over it.

During the New Kingdom, it became standard for men to wear a longer skirt that covered a shorter one, often pleated from the hip. The torso was covered with a kind of tunic with gaps at the sides to insert the arms, although there were also tunics with sleeves and a gap at the top to insert the head. Wigs were a fundamental feature of both men's and women's clothing, as both sexes shaved their heads. Wigs were made from natural hair and vegetable fibers. Probably on the grounds of hygiene, as the proliferation of insects was a sign of plague, Egyptians would systematically shave their heads and their entire bodies, using blades and tweezers. Bathing and cleaning their fingernails and mouths were fundamental to the daily hygiene of the Egyptian people.

Despite being renowned as a hardworking people, Egyptians enjoyed free time and spent it doing what they enjoyed most. Their favorite hobbies can be seen in numerous paintings in temples and tombs. Dancing, sport, and board games aside, which were hobbies shared by all strata of Egyptian society, nomarchs and courtiers in general were fans of hunting and, surrounded by a large entourage of servants and slaves, they ventured into swampland or desert regions to hunt at their pleasure. They also enjoyed fishing, either on the banks of the river or on boats.

Wigs, made from natural fibers, were a fundamental feature of both men's and women's clothing, as both sexes shaved their heads

WOMAN WITH JAR
This woman is carrying an ointment jar and a sack of lily petals, presumably for part of a celebration.

The City of Thebes

Imposing cities, such as Memphis and Thebes, were built on the Nile as a testament to the Pharaoh's power and devotion to the gods. A religious, political, and trade center for almost 1,000 years, Thebes (present-day Luxor) was the most important of all.

Thebes, a holy center

The Greek poet Homer called it "the city of the hundred gates." It was the capital of the empire and the spiritual and political center between 2055 BCE and 1070 BCE during the Middle and New Kingdoms. Despite the official capital being relocated later, Thebes remained the most important holy city in Egypt.

TRADE ROUTES
Using the Coptos caravan route, goods and slaves from the Persian Gulf or the Red Sea arrived in Thebes.

The structure

Thebes, like other Egyptian cities, stretched out along the banks of the river. Its main avenue ran parallel to the course of the river and countless streets converged on its central square, around which ordinary members of society lived their lives. This contrasted with the lifestyle of the nobility and the clergy, who stayed in the palaces and temples. Civil constructions were made from adobe and brick.

EASTERN DISTRICT
The center of the city's lifeblood. It housed royal palaces, temples, and administrative buildings.

THE TEMPLE OF AMUN IN LUXOR
Erected around 1390 BCE by Amenhotep, the architect appointed by Amenhotep III, it was later expanded by Ramesses II. Located at the heart of Thebes, it is one of the best preserved urban constructions of Ancient Egypt.

AVENUE OF THE SPHINXES
It joined the Temple of Luxor to the Temple of Karnak, almost 2 miles/3 km away. It is believed to have been flanked by 1,350 sphinxes.

DYNAMISM
The city was very active, with a district of artisans, a port at which products arrived, and a bustling market that provided the population with sustenance every day.

LOCATION
One of the main sites of Ancient Egypt, Thebes was the southernmost city. It is believed that at its peak, over 650,000 people lived there.

THE NILE
According to Homer, it was thanks to the river that Thebes accumulated wealth, only surpassed by the grains of sand in the surrounding area.

AFRICAN WEALTH
Some caravans traveled the banks of the Nile from Punt, on the African coast of the Indian Ocean, or over routes through the Sahara Desert.

Egyptian Housing

Egyptian houses were made from adobe bricks and they housed the entire family. These humble homes were just one story, small in size, and featured four different rooms. Their structure has been determined thanks to the excavations at Deir el-Medina and the working class district of Tell el-Amarna.

Security

As the windows contained no glass, stone grilles or rails were installed to prevent strangers or animals entering the house. Valuable objects were usually stored in the basement, the most difficult place to access.

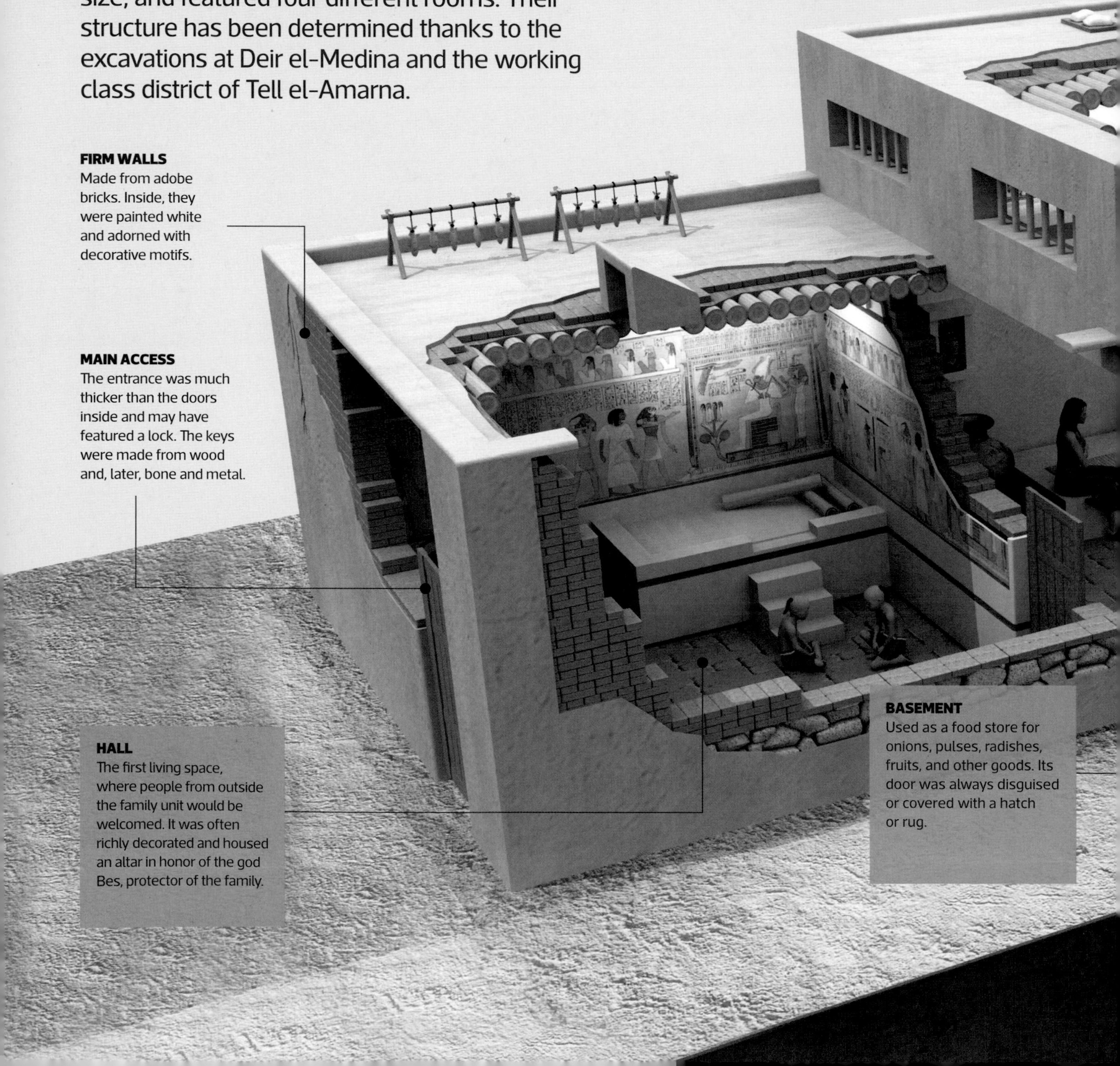

FIRM WALLS
Made from adobe bricks. Inside, they were painted white and adorned with decorative motifs.

MAIN ACCESS
The entrance was much thicker than the doors inside and may have featured a lock. The keys were made from wood and, later, bone and metal.

HALL
The first living space, where people from outside the family unit would be welcomed. It was often richly decorated and housed an altar in honor of the god Bes, protector of the family.

BASEMENT
Used as a food store for onions, pulses, radishes, fruits, and other goods. Its door was always disguised or covered with a hatch or rug.

THE ROOF
The Egyptians dedicated most of their day to work. The roofs served to dry and salt the meat and fish they consumed.

VENTILATION
Windows were positioned in the upper part of walls, to prevent direct sunlight and sand from entering. They were small, rectangular openings.

COT
Beds were simple mats laid flat on the floor, or cots made from woven hemp.

KITCHEN
Kitchens were well-equipped, with spaces for storing cutlery and cooking utensils. They featured mortars and original clay ovens where they made, among other dishes, leavened bread.

FAMILY SPACE AND FURNITURE
In the central room, members of the family would meet daily. It included benches and may also have contained stools, tables, and ceramic vases. The best furniture was made from wood, and had carvings and paintings.

Residence of the nobility

The most powerful Egyptians built their houses with spacious gardens and patios with plants, palm trees, and flowers. They even built pools to cool off and escape the sweltering heat of the desert. Some of the houses had an internal courtyard at the center of the house, which offered light and shelter from the intense sun.

Domestic Furniture

Generally, little domestic furniture was used; only the upper classes were able to afford large amounts of furniture, which, rather than serving a useful purpose, indicated their social status. Seats, stools, a range of different-sized chests, beds, and headrests were the most frequently owned pieces of furniture among rich families in Ancient Egypt.

CHESTS
Used to store objects. They had four legs and were usually cube-shaped, although some chests had curved lids. They were ornately decorated with appliqués and inlay.

Materials

Wood was the most commonly used material in furniture production. Generally, in particular from the New Kingdom onward, furniture was well-crafted and featured ivory inlays, semiprecious stones, or fine wood covered in gold or paint. Parts were put together using joints, wooden pins, or nails.

STOOLS
They were either fixed or folding. Folding stools often had a woven vegetable fiber or leather seat.

DECORATION
Decorative motifs were commonplace. Many decorations had symbolic meanings.

Reliefs and paintings

In addition to the furniture found in funerary chambers, reliefs and paintings found on their walls have provided more information about Egyptian furniture, as it often appears in the scenes depicting daily life. In this image, two noblemen are sitting on chairs, playing *senet*, a popular board game.

A furnitureless existence

More humble homes featured only long benches built into the wall, made from the same material as the structure of the house, and they served as a seat, table, bed, or sideboard. Generally, the house had no other furniture, with the exception of the odd stool or container to store water and foodstuffs.

REGAL
Seats had a backrest and some featured armrests; given their robustness and beauty, they looked like genuine thrones.

LEGS
The legs on seats, stools, and beds often took on zoomorphic shapes (often imitating the legs of a lion or bull).

Furniture preserved to this day

Certain items of furniture found in tombs have remained intact, such as in the tombs of Hetepheres, wife of Sneferu (fourth Dynasty), and Tutankhamun (eighteenth Dynasty). The time that separates them, almost 2,500 years, has made it possible to establish that the style of furniture barely changed at all, and that the most important differences are attributable to better tools.

Royal furniture. Located in Giza, remnants of several pieces of furniture, such as a bed, a throne, and various chests, were found in the tomb of Hetepheres.

HEADRESTS
They served to support the head when somebody stretched out on a bed. They were made from wood, ivory, or metal.

BEDS
Beds were a genuine luxury and scarcely found. They featured a single footboard and four zoomorphic legs. They were slightly inclined, meaning that the user's feet were lower than his/her head.

Family

The family represented the basic social unit in Ancient Egypt. The title of "lady of the house," assigned to married women, demonstrates their importance in the domestic setting. Children were the main objective of marriage, which was a voluntary and private matter between the couple.

Marriage

The father of the bride was responsible for choosing a spouse for his daughter, although she would have the last word. Members of the same social class married, and it often involved the coming together of two blood relatives. Among royalty, heirs to the throne could marry their sister to safeguard the continuation of the dynasty. Slaves were not allowed to marry.

Pregnancy and maternity

The most dangerous moments in a woman's life were during pregnancy and giving birth. Infant mortality rates were very high until the age of five. During birth, which took place at home, the woman was assisted by one or more midwives, and the use of an amulet was commonplace to ensure everything went well.

Playtime. Egyptian children played with toys made from wood or natural fibers. This image shows a wooden doll and a horse on wheels.

LEISURE TIME
Board games were popular among Egyptian noblemen as well as members of the working class.

Divorce

Like any other contract, if a marriage did not work out, it could be terminated. The main causes of divorce were infertility and infidelity; the latter was considered more serious if the woman was responsible. Among the common people, relationships were monogamous.

The Dwarf Seneb and His Family. Sculpture from the fourth Dynasty carved from limestone.

CHILDREN
All children were well looked after, regardless of their gender. Only boys were allowed to attend school.

CARING FOR THE HOME
Women were responsible for raising children and keeping the house, while their husbands worked outside the home.

CARING FOR THE ELDERLY
Children most often took care of their parents when they grew old. If a woman was widowed, her children took care of her.

The Role of Women

The main role of an Egyptian woman was as mother and wife. Her social position was determined by her father and husband. Although she lived her life in the background, her circumstances were better than those of women in many other ancient civilizations: she was equal to male counterparts before the law, could own property, manage income, and receive inheritances.

CARE AND HYGIENE
Personal hygiene was of the utmost importance to Egyptian women. Rich women cared for their skin using hydrating oils.

Domestic life

The responsibilities of peasant women were to take care of their husbands, bear offspring and raise them, in addition to running the household and working in the fields, if necessary. They could also trade the crops they cultivated as well as the handicrafts they made.

Working life

Women were excluded from many jobs, although others were occupied almost exclusively by women: they could be musicians and dancers, weavers, servants, paid mourners, and midwives. They could also work as peasants, officials, or as part of the priesthood.

FERTILITY
The importance of female fertility can be seen in the multitude of images found. These small statues formed part of a domestic cult and were responsible for safeguarding conception and a successful birth.

SHAME
Toward the end of the empire, with the concept of feminine shame, women progressively started to cover their bodies.

BEAUTY ENHANCEMENTS
Necklaces, headdresses, rings, and bracelets were commonly worn by Egyptian women.

SERVANTS
The richest families had servants who were tasked exclusively with washing their clothes and keeping their wigs in good condition.

Physical appearance

In paintings and reliefs, the female body tended to be depicted as slim and well maintained. Beauty and youthfulness represented the feminine ideal, with an emphasis on the hips and breasts, parts of the body related to maternity.

TABOO
Menstruation, considered impure, was a great taboo among women. On days of menstruation, both women and their husbands may have missed work.

Royal women

Although in principle women were unable to hold positions of power, wives, mothers, and daughters of the Pharaoh enjoyed great influence and prestige. A mother could be regent while her son was underage. Some wives of Pharaohs have gone down in history, such as Nefertiti, the wife of Akhenaten, who participated in the reforms proposed by her husband. Nefertari, wife of Ramesses II, was honored in the construction of a temple, along with the goddess Hathor.

PHARAOH QUEEN
Hatshepsut, daughter of Thutmose I, ascended the throne on behalf of her stepson, Thutmose III, after the death of her husband. Shortly afterward, she was proclaimed Pharaoh and stole power away from her stepson. During her reign, which contributed to the splendor of the eighteenth Dynasty, she established new trade routes, organized several military campaigns, and erected a huge funerary temple.

Children's Education

During infancy, mothers cared for their children. Later, the child's father was responsible for its education. From the Middle Kingdom onward, evidence of schooling, usually dedicated to teaching children how to read, was documented.

Schools

At elementary school, the concepts of reading, writing, and mathematics were taught. These were followed by geometry, astronomy, and gymnastics. Students attended high school until the age of 17, and would often go on to occupy positions as scribes and state officials. Although open to everyone, only the children of the wealthiest members of society could attend school, as the offspring of poorer families had to work.

Educating young girls

Young girls received less education than young boys. Basically, they were shown how to run a household. In the upper classes, they were also taught music and dance and occasionally to read and write.

Running the school

The basic tenets of elementary school, where a teacher taught with the children sitting around him, were:

DISCIPLINE
Generally, schools employed a very strict form of discipline, in which punishment, smacks, and lashing were not uncommon.

CALLIGRAPHY
Children were first taught to read and write. Stone slates were used for writing, which were much cheaper than papyrus.

METHODS
Reading, memorizing, and transcribing texts held the keys to pursuing more specialized disciplines.

The House of Life

This was the name given to specialist schools, which were veritable temples of wisdom and only accessible to the wisest men in the country. Knowledge of several sciences and arts was imparted here, ranging from theology and politics to law and ritual magic.

FUTURE OFFICIALS
In order to become an official, reading, writing, counting, mathematics, geometry, and engineering were essential.

TEACHERS
Scribes were responsible for education in schools.

Clothing

Egyptian clothing was designed to counteract the region's hot climate. White linen, worn in the form of long tunics by women and short skirts by men, was the most commonly used material. The design of the dress remained almost unchanged throughout the empire.

WIGS
Used by both men and women, they protected the head from the sun. They could be made from woven black linen or natural hair, fastened using beeswax.

Garments

Clothing was a reflection of a person's socioeconomic level. The basic garments worn by men and women were common to all levels of society; what distinguished them was the quality of the fabric. Other garments, such as tunics, coats, and shawls, were typical among the upper classes, to whom the robes were not bothersome as they did not participate in physical chores.

TORSO
Men could leave their chests bare, or wear a tunic with wide sleeves and a shawl across their shoulders.

NUDITY
Children and slaves often wore no clothing.

SKIRT
Made from white linen fastened at the waist, this item of clothing was usually worn by men. During the Old Kingdom, they were knee length, while during the Middle Kingdom they were calf length and in the New Kingdom they were accompanied by a long pleated skirt.

Cleanliness and personal appearance

Egyptians took great care of their hygiene and personal appearance. Daily hygiene involved cleaning their toenails and washing their mouth. Men and women shaved their heads and wore wigs. Makeup, perfumes, and jewelry complemented a style of dressing characterized by its simplicity.

ADORNMENTS
Eye makeup, wigs, and jewelry were used by both men and women to decorate their bodies.

ELEGANCE
During the New Kingdom, women's clothing became more elegant and refined. Blouses, fitted to the body, left their right shoulder exposed.

LONG LINEN DRESS
The most common garment among Egyptian women—normally, a pleated, white, semitransparent dress that made it possible to see the wearer's figure.

SANDALS
Egyptians often went barefoot. When sandals were used, they were made of papyrus when worn by humble classes, and leather when worn by the well-to-do.

Jewelry

Egyptian jewelry is distinguished by its beauty and perfection. The wearing of jewelry was common among both men and women and, decorative value aside, it was considered protective and magical. Jewelry was also an indication of a person's social standing or job; some pieces were used in rituals and others adorned temples and palaces.

Materials

Gold, copper, and silver were the most commonly used metals in jewelry production. Ostrich eggshells, turtle shells, horns, bones, and ivory were also used during the first years of the Old Kingdom. Later, the use of precious and semiprecious stones was introduced, such as agate, amethyst, emerald, feldspar, lapis lazuli, malachite, obsidian, and others. Each color was designated a magical property.

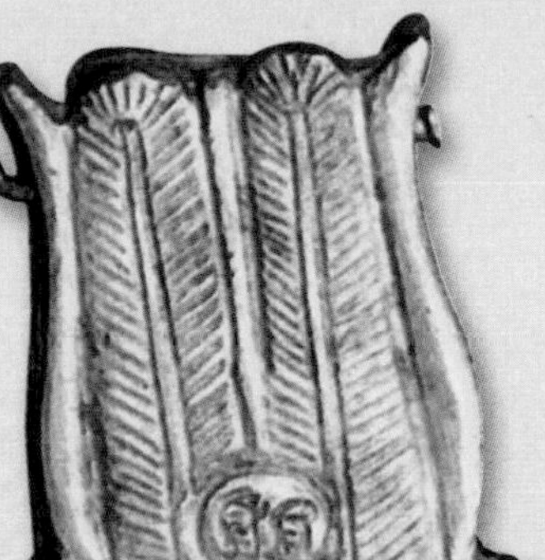

GOLD
The most precious metal and most used in the production of jewelry. It was obtained from gold mines in the Nubian desert.

AMULETS
This amulet is one of the pieces that formed the burial paraphernalia of Tutankhamun, together with other jewels.

Full of symbolism

Beetles, scorpions, the moon, the sun, and the eye of Horus were common motifs in Egyptian jewelry. In addition to decorating the person wearing them, they were talismans or amulets that served to ward off evil, both in physical form and those resulting from spells or evil eyes.

Eternal jewelry

When a member of the nobility or the royal family died, the most important objects and jewels that had formed part of his/her life were included in the burial. Other pieces were also made that were only to be used in the afterlife.

COLORS
The colors of ornaments that adorned the jewels gave them a special meaning.

Jewelry techniques

Many modern jewelry techniques have been inherited from ancient times. To create such magnificent pieces, jewelers used molds. They soldered them, fused pieces of glass and stone, cut and embossed them with a chisel, carved out engravings, and used granulation and filigree techniques. They polished the pieces by rubbing them with sand.

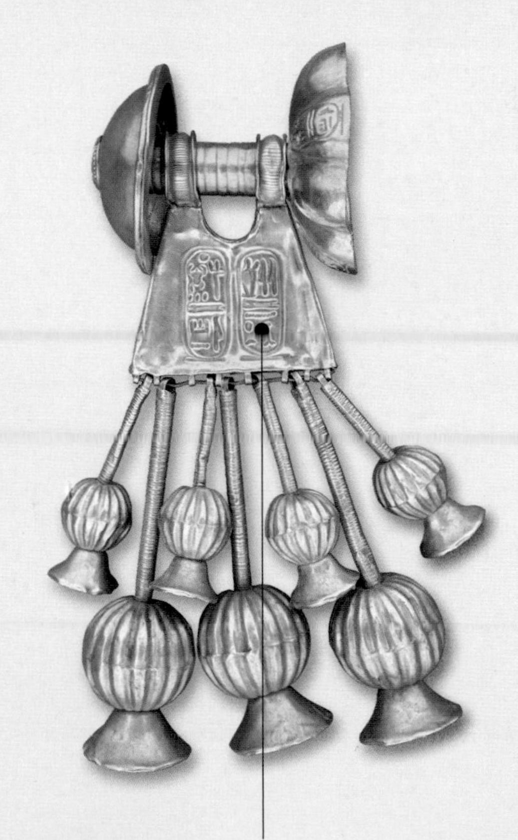

NECKLACES AND BRACELETS
The most common forms of jewelry were bracelets, pectorals, armlets, anklets, and rings.

MILITARY MEANING
The fly necklace was an important form of recognition in the Egyptian military. This one was presented to Queen Ahhotep who, during the period of coregency with her son, Kamose, achieved notable victories and was a true leader.

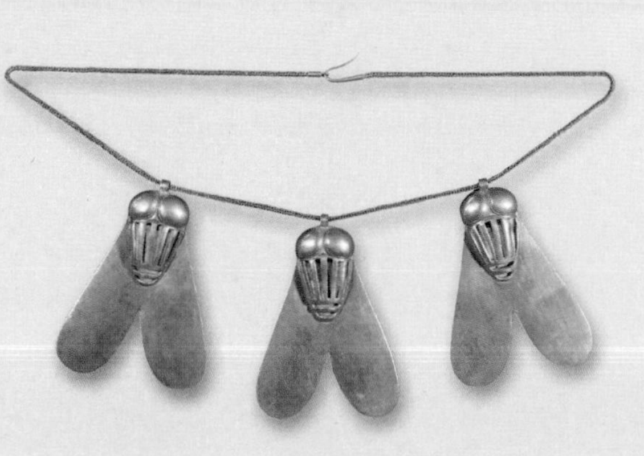

ROYAL SEALS
Many ornaments were decorated with the Pharaoh's cartouche. They could be pieces that belonged to the Pharaoh himself, or to another member of the royal family. Alternatively, they could represent a gift to the Pharaoh.

Food

Bread, beer, bean and lentil stews, milk, cheese, and fish from the Nile formed the staple diet of the common people. Meat, rarely consumed in humble homes, was common at the dining tables of the Pharaoh, the court, and powerful families. Other delicacies would only be consumed on special occasions.

Basic foodstuffs

Cereals, such as wheat and barley, were the staples of agriculture and were predominant in the Egyptian diet. They also grew and consumed pulses and vegetables such as celery, watercress, asparagus, lettuce, garlic, onions, chickpeas, and lentils. The most popular fruits were dates, figs, grapes, apples, and melon.

LIVESTOCK
As an activity, rearing livestock was closely linked to agriculture: reared animals helped with chores in the countryside and served as a source of food. Egyptians also kept pets.

BREAD AND BEER
Produced on a daily basis at home. Beer, made by fermenting barley flour, was considered a highly nutritious source of food as well as a drink. It was often sweetened using honey, dates, or spices. Flour was also used to make desserts with honey and fruits.

FISH
Although fish was prohibited for priests responsible for leading certain ceremonies, it was a basic source of food for common people. The most common catch, caught using nets, were catfish and bass from the Nile. The fish were dried or smoked to preserve them.

Select products

Some foodstuffs were not within the reach of all sections of society. Generally, meat was consumed less often than fish and certain sources of food were considered more select than others. On the other hand, wine was frequently consumed at banquets put on by high society.

CONSUMPTION OF MEAT
The most common meats at Egyptian tables were lamb, beef, and game. Cuts of beef, geese, and duck were reserved for certain banquets. When a Pharaoh died, select cuts of meat were mummified so that he could consume them in the afterlife.

WINE PRODUCTION
Initially, Egypt imported wine from Palestine until it mastered the art of production, around 3100 BCE. Grapes were collected by hand, then crushed and fermented in jars. Each jar bore a label with the year of harvest, the origin of the wine, and the winemaker.

Making Bread

INDOOR CHORES
Grain was processed in dry environments, sheltered from sunlight.

The main crops sown by the Egyptians along the banks of the Nile were cereals. Wheat, along with barley (used to make beer), was the most important. It was used to make baked bread, the staple food of most of the population.

Taxing foodstuffs

The Pharaoh's granaries were distributed across the country. Specialist officials organized the collection and storage of harvested cereals. The state paid quarry workers and those laboring on large-scale construction projects with the cereals collected, in addition to bread and beer.

DISTRIBUTION
Following the harvest, one part of the grain was designated for paying taxes, another was reserved for the following sowing season, and the last part was separated for grinding.

This wooden model from the Middle Kingdom depicts the steps required to produce bread.

MIXING
Moistened flour was mixed with yeast. It was kneaded in a wicker basket until the required consistency was achieved and then placed in the oven.

WATER
Stored in stone wells or hauled from the banks of the river every day.

OUTDOOR CHORES
Unlike the upper echelons of society, who installed their oven in a courtyard, most people cooked on the rooftops.

BAKED
Bakers used firewood or dry animal dung as fuel.

GRINDING
Women tended to dedicate themselves exclusively to this task. They ground the grain patiently until they obtained the consistency of flour required to bake bread.

NUTRITIONAL BREAD
Thanks to hieroglyphic paintings, it has been established that Egyptians mixed dough with the seeds from different plants in order to make bread more nutritious.

Trade

In a society in which most people made a living from agriculture, trade was limited. Commerce was based on exchange, or bartering, as it was an efficient method, given that the basic needs of the people were mostly the same; grain and oil often served as currency.

Location

The preferred site for holding the market was the open space around the docks, given that all Egyptians lived near the Nile. Wives of peasants sold clothing, grain, or game. The proximity of the river also made it possible to sell goods to sailors on their boats.

Trade tools

The population needed very little: grain to make bread and ferment beer, dried fish, vegetables, fruit, meat and game, linen to make cloth, and mud bricks for building houses. Small luxury crafted items could be exchanged in the event of a surplus when a family had already been able to cover all its basic needs.

MEASUREMENTS
Containers featured no form of standardized weight or size measurements. This issue led to the adoption of precious metals as currency.

SCALES
Enabled an equivalence in the weight of products to be established, but not the value of goods. This difficulty required creating fixed rates of exchange.

TRANSPORT
Few people had a cart and draft animal to take produce to the market. Thus, individual bartering was monopolized by the rich, who owned their own methods of transport.

GIFTS AND SOCIAL STANDING
Among ancient agrarian societies, exchanging goods served not only an economic purpose, but also had a social significance. Exchanging gifts denoted social standing and ensured the bond was maintained.

EXCHANGE
Even after introducing minted coins, bartering remained the most common form of trade among country dwellers for centuries.

MEATS
The meat trade was significantly limited by the challenges of preserving it in such a hot environment. Salt, used for preserving, was also very expensive.

Foreign trade

International trade relations were almost a monopoly exclusive to the Pharaoh. Egypt had a great wealth of resources, but had to import timber, iron, silver, tin, and lead. The main routes were: Egyptian-Phoenician, by means of which Egyptian products reached the central and eastern Mediterranean and Mesopotamia; the Red Sea, to import produce from Africa and Arabia; and the route of the oases, which was another important trade route with Africa.

Detail of a relief featuring a trading boat.

Leisure and Fun

In their spare time, Ancient Egyptians enjoyed various activities. Music and dance, hosting banquets, sporting competitions, and board games were all among their favorite hobbies.

Board games

Very popular in Ancient Egypt, the most commonly played was *senet*. Twelve pawns were positioned on a board with thirty squares. The pawns were then moved from square to square depending on the outcome of tossing sticks. The player who arrived at the end first was declared the winner.

Senet
Reconstructed model based on a real board.

Music

Accompanied rituals and prayer, as well as festivals and processions. The harp was the most common instrument, although it was often accompanied by a three-string guitar. The *sistrum* was also very popular; this percussion instrument had a U-shaped wooden frame, with a handheld shaft. Upper-class women were often well versed in music.

Eroticism and sexuality

Sexual activities were enshrouded in a primarily religious connotation, as they were associated with fertility and reproduction rites. However, Egyptians also enjoyed significant freedom in terms of sexual pleasure. There are many examples of erotic scenes in funerary wall paintings, ranging from the sexual act itself to simple nudity, especially among women.

DANCE
At banquets, music and dancing entertained guests while the food was prepared and served.

FOOD AND DRINK
Wine, beer, and meat were all served in abundance at banquets. Beef was reserved for this type of celebration.

Sporting challenges

Races, high jump, javelin, archery, boxing, and wrestling were the most commonly practiced sports in Ancient Egypt. In addition to being practiced for leisure, many sporting challenges had a ritual meaning. Another sport practiced by both men and women was swimming.

Wrestling. Wrestling scenes are common in the wall paintings of the Pharaohs' tombs.

Banquets

Frequently held by members of the royal family, they served to celebrate significant occasions, such as births, marriages, and religious festivals. Music, dance, and acrobatics were common at such feasts.

OFFICIAL BANQUETS
The Pharaoh often organized banquets, to which he invited prestigious members of the empire.

Hunting

Very common during the Predynastic period, hunting large animals for subsistence lost importance with the evolution of agriculture and rearing livestock. Those who continued to hunt for subsistence did so using nets and traps. Among the upper classes, the harpoon was used in recreational hunting.

In the desert

The common gazelle, of which there were many on the Nile flatlands, was easy prey for desert hunters. Addax (antelopes), oryx, jackals, hyenas, and ostrich were also hunted.

In the river

The Nile was the source of life not only for its floods, which helped agriculture, but also because of the animals that inhabited it. There, hippopotamuses were hunted. Crocodiles and turtles were also captured and mussels were harvested. Nets were used for fishing.

HUNTING GAME
Ducks, geese, herons, cranes and quail were the most common types of game in the wetlands. They were hunted both recreationally and by those who needed them for food.

Hunting and fishing scene from a Theban tomb, dated to 1400–1390 BCE.

RECREATIONAL HUNTING
This member of the nobility can be seen enjoying leisure time, hunting along the banks of the river.

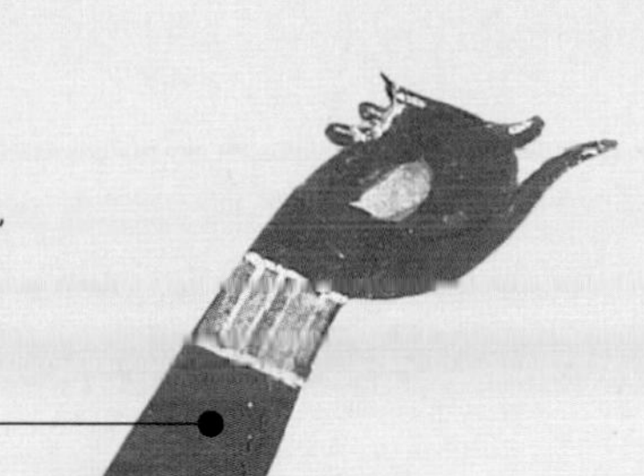

FAMILY
In this image, the hunter can be seen hunting game accompanied by all the members of his family and his servants.

Weaponry

The Egyptians used bows and arrows, spears, lassos, and dogs to hunt large prey, and nets and traps for smaller prey. Hunters often traveled on foot, although there is also evidence of hunters having traveled by chariot.

Ceremonial shield This shield was part of the treasure found in the tomb of Tutankhamun and depicts a scene hunting lions.

Hunting hippos

Hunting hippos was a difficult task and required large amounts of effort and tenacity. From their boat, the hunters thrust harpoons into the animal and followed it as it fled until it bled to death.

In this bas-relief found in a tomb at Saqqara, various hunters can be seen in their boats trying to capture three hippos on the Nile using spears.

Inventions and Ingenuity

From the time of the Old Kingdom, in the third millennium BCE, Egyptian civilization developed a whole range of systems for optimizing navigation and trade, tools for increasing agricultural production, and procedures for making administration more efficient.

The Lighthouse of Alexandria

Built in the third century BCE under the Ptolemaic Dynasty, an offshoot of the Greek monarchy that ruled Egypt for almost three centuries. Considered one of the Seven Wonders of the Ancient World, it was destroyed by an earthquake in the fourteenth century.

Height 400 ft/ 12 m, the highest tower of the time.

Light

Projected up to 30 miles/50 km to guide trade boats.

MATERIALS
Stone covered with white marble.

RAMPS
Used for access purposes, they rose up in a spiral shape.

Swape or shadoof

First used during the New Kingdom, in 1550 BCE, this formed an essential part of the irrigation system for Egyptian crops.

FUNCTIONALITY
To retrieve water, a bucket was submerged in the river and released when it was full, by means of a system using levers and counterweights.

COUNTERWEIGHT
Stones bound to one end of the lever enabled the receptacles containing water to be raised.

WATER DISTRIBUTION
Water was distributed from the receptacles to the canals, then dispersed onto the fields.

LOADING WATER
Using buckets that hung from the end of the levers.

MIRROR
Reflected sunlight during the day and the light of the fire at night.

POSEIDON
The identity of the statue on the dome of the lighthouse is subject to debate. Many support the theory it was the god of the seas—Poseidon.

BONFIRE
Lit using fuel that rose up through the central shaft.

TRITONS
In Greek mythology, Triton is the messenger of the seas.

PHAROS ISLAND
Many languages have named this construction after this island.

Papyrus paper

One of the main Egyptian contributions to the world. It was used for writing on and its production was a royal monopoly. The oldest fragment was found in the necropolis of Saqqara and belonged to the era of Den (first Dynasty).

1. First, strips were cut from the stem of the papyrus plant.

2. The strips were placed horizontally and vertically before they were pressed, to extract the sap.

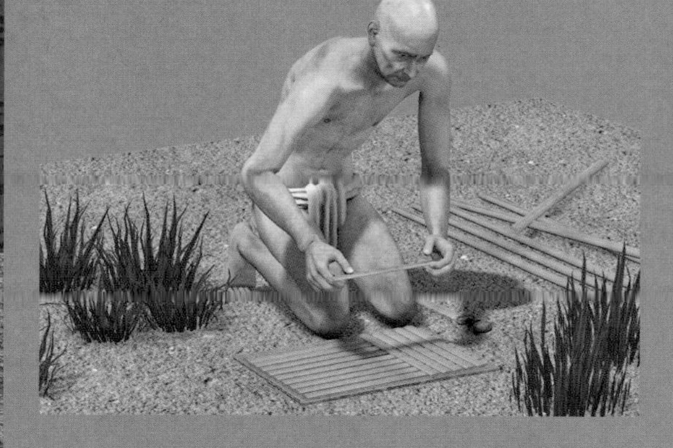

3. Later, they were crushed with a piece of ivory over a number of days, until the papyrus was ready for use.

Medicine

Medicine was a more advanced branch of education that scribes studied in later years. Although some components involved in treatment were more related to magic than pharmacopoeia, Egyptian medical practice was based on science.

Serqet

The gods also played a key role when it came to healing. Invocations to Serqet, the scorpion goddess, were one of the most common ways of preventing, or curing, scorpion stings.

The circumstances of Egyptian doctors

Employed by the state and financed by the Pharaohs, they had no need to charge patients. They commanded the utmost respect and spent their days either in research or serving the population. Through observation and accumulating experience, Egyptian doctors reached quite accurate solutions when treating their patients' illnesses.

Surgery

It is known that the Egyptians practiced surgery, applying adhesive strips to suture wounds.

ASSISTANTS
They helped by keeping the patient still and, where necessary, helped with medical instruments.

DOCTOR
Those in the profession had a good reputation, to the extent that their services were called upon outside the empire's boundaries.

Egyptian knowledge

Several papyri have been preserved that served as teaching manuals as well as reference. Most contained treatments for various illnesses. They also feature data on the structure of the body, its organs and functions. Some illnesses for which the corresponding treatment appears on these papyri include women's illnesses, ear and stomach problems, and heart, lung, or liver issues.

Circumcision

One of the most mentioned and featured medical practices in the texts. Practiced by the Egyptians since ancient times, this took place when a boy entered manhood and was more than likely related to puberty rites.

PHARMACOPOEIA
Remedies that contained medicinal properties were preserved on papyrus, including their preparation. Plants like garlic, poppies, acacia or castor leaves, and cereals were used to produce medicines.

HEALER
Responsible for producing magic spells considered necessary to ensure the treatment was effective.

Mathematics

Egyptian mathematical knowledge is understood thanks to a series of papyri used to teach the scribes. These demonstrated operations for practical purposes and suggested that mathematical calculations were all based on addition and subtraction.

Pharaonic calculations

The papyrus of Ahmes, or Rhind (1650 BCE), is the best known source about Egyptian mathematics. It comprises 87 problems and solutions, with exercises in arithmetic and algebra, area calculation, and the concepts of trigonometry. "Precise calculation, to attain knowledge of everything that exists and all the dark secrets and mysteries," is the first sentence of the text.

First, the practicalities

Egyptians seem to have had an interest in practical arithmetic, which enabled them to make calculations such as the harvest obtained based on the cultivated surface area, labor, and the time required for a given task, administration of the temples, the area and volume of different figures, and so on.

PLUMB BOB
Made it possible to establish whether roofs or floors were horizontal and to trace a line from north to south.

SLOPE CALCULATIONS
Establishing the *seqt*, or slope, of a flat, inclined surface was important to architects.

Wisdom of Pythagoras and Euclid

Greek philosopher Pythagoras, proponent of the theorem on right-angled triangles, spent 21 years studying with Egyptian mathematicians, priests, and architects. Euclid, a founding father of geometry, also received his education in Alexandria, Egypt.

HIEROGLYPHIC DECIMAL SYSTEM
There were different symbols for representing the numbers 1, 10, 100, 1,000, and 10,000, but there was no symbol to represent 0.

Sacred Triangle

The name given to a right-angled triangle (the easiest to construct) by the Egyptians; its sides have lengths of 3, 4, and 5 or maintain that proportion. It may have been used to obtain right angles for constructions, and held a strong symbolic significance.

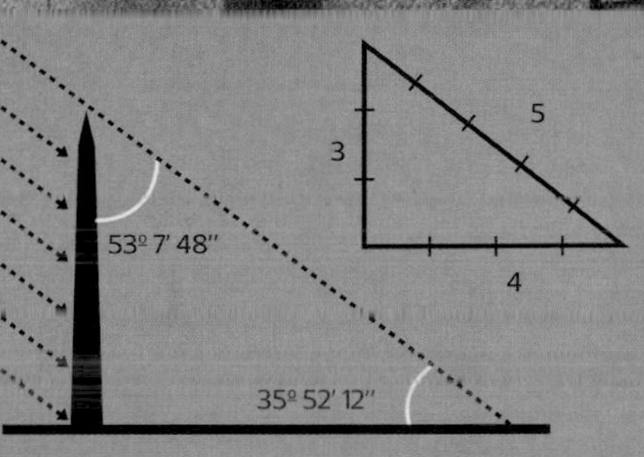

INTERMEDIATE NUMBERS
Intermediate numbers were written as multiples of a simple number. If 1 was one stroke, 8 was eight equal strokes.

ON WALLS
Numbers carved into the walls of temples usually expressed the number of people killed or prisoners taken in combat.

4

CHAPTER

MYTHOLOGY AND RELIGION

THE GODS AND THE AFTERLIFE

Ancient Egyptians elaborated myths about the creation of the world and the solar cycle and worshipped a number of deities. Death was a central concern of the population, and the journey of the deceased to the afterlife was guaranteed by a series of rituals.

The religious world of the Ancient Egyptians, given the number of beliefs involved and its extension throughout time, was extremely complex. Polytheism, the official religion, like all beliefs, tried to respond to all of life's questions, especially about death. Unlike in other civilizations, such as Ancient Greece, Egyptian religion did not involve mythical theories about the evolution of the gods, given that their deities, based on the theocratic nature of Ancient Egyptian society, tended to take on an absolute and unchangeable form. Gods and goddesses featured in the pantheon had always been, and would always continue to be, just as they were.

Conversely, myths did exist about the great processes of life, such as the creation of the world. In this particular case, Ancient Egyptians believed that the god Re, also known as Re-Harakhty or Atum, was the creator. The most popular myth saw him as the creator of the universe as a result of his escape from chaos. After setting foot on a mound that emerged from the water, he created two further gods using the soil, Shu and Tefnut, who in turn gave birth to Geb, god of the earth, and Nut, goddess of the sky. Finally, the children of Geb and Nut—Osiris, Isis, Set, and Nephthys—completed the creation of other beings and elements of the world. Other myths, mainly recompiled by Roman historian Plutarch, refer to dramatic episodes, such as the murder of Osiris by Set, the conception of Horus by his mother Isis with the deceased body of Osiris, and the definitive defeat of Set at the hands of Horus.

The solar cycle was the source of numerous myths which, in reality, represented variations of the same series of topics, such as: the sunrise, its journey by boat through the heavens; its decline, related to the process

MUMMY MASK
Preserving the maximum integrity of the corpse ensured the soul made a successful journey into the afterlife.

SEKHMET
Statue dated by archaeologists to between 1390 and 1353 BCE. Sekhmet was a warrior deity who protected the sovereign.

of growing older in life; and the sunset, as a symbol of death. Some myths contradicted others: while, in the creation myth, Nut is a granddaughter of the sun, in the solar cycle myths, Nut became its mother. At dawn, the sun leaves her mouth and, at dusk, it retreats back inside. In some mythical variations, Nut was replaced by the goddess Hathor, for which the solar cycle was linked to rebirth. Other myths addressed the end of humanity as an inescapable event, as a result of which Re and Isis would give birth to new beings that would start a new life cycle on earth.

Attributes of the gods

The various deities maintained a special relationship with each aspect of reality: for example, Re was associated with the sun; Hathor, with women; and Ptah, with craftsmen. Meanwhile, each god or goddess had a whole network of attributes that went beyond his or her specific purpose: Re was also associated with preventing droughts; Hathor, with cosmetics, and Ptah, with the street trading of various products.

In his journey across the heavenly sky, the god of the sun was accompanied on his boat by a crew of minor gods, who often personified his other attributes. At a given point in its course, when the strength of the sun dwindled, the boat was dragged away by a pack of jackals. When the sun god rose from the horizon, all of creation rejoiced. He was welcomed by numerous gods and goddesses, in a ceremony presided over by the Pharaoh. As the rest of society was banned from these festivals, the Pharaoh, as a representative of both the human and the divine, served as an intermediary between mere mortals and the gods, which reinforced his power at the top of the empire's social structure.

AMUN
Zoomorphic image in a wall painting from the fourteenth century BCE, featuring the ram's head of the god Amun, who, in association with Re, was considered the creator of the universe.

Regional variants

The organization of the gods was different in the various regions of Egypt. In general, they were distributed into triads made up of two major gods and one minor god. The imposing triad of Thebes comprises Amun-Re, Mut, and Khonsu, whose temples can be found in the sacred complex at Karnak. The three deities represented the typical family structure of father, mother, and son, and the magical connotations linked to the number "three."

In Heliopolis, the sun god was maintained independently. He was given two wives, Iusasset and Hathor-Nebethepet, who personified sexual elements belonging to the creation myth. In Memphis, the association of the four main gods, Ptah, Sekhmet, Nefertem, and Seker, differed: the first three comprised the triad, while Seker, god of the city's necropolis, was assimilated as the figure of Ptah. Hathor and Neith, whose cults were important in Memphis, were excluded from this main group.

In both the official and local pantheons, the gods could be linked and identified in different ways. A deity could be given many names, some of which were assumed, based on the existence of previous deities in a given place that were still maintained in popular imagination. As a result, in certain regions, Amun-Re was known as Amun in his appearance as Re, and in other places, could even be expanded to Amun-Re-Atum. Generally, the cult of Re predominated throughout Egypt, as it formed part of the universal nature of the solar cult. The case of the god Osiris, on the other hand, was entirely different: in Abydos, he was known as Khenti-Amentiu, linked to a local god from an ancient cult belonging to Lower Egypt.

All the gods mentioned, considered major gods, existed side by side with other deities that were only the object of worship in a very limited number of

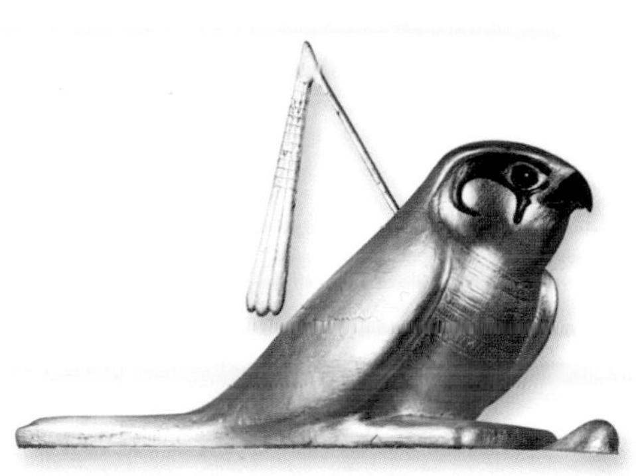

GEMEHSU
Depiction of the falcon god, primarily associated with Horus. In statuettes, like this one made from solid gold found in the tomb of Tutankhamun, it constituted a funerary offering.

ANUBIS
Represented with the head or entire body of a jackal, Anubis accompanied the deceased on their journey to the afterlife and watched over the scales during the Judgment of Osiris.

Generally, the cult of Re, the god of the sun, predominated throughout the lands of Egypt, as it formed part of the universal nature of the solar cult

FUNERARY CHEST
Dated to the fourteenth century BCE, designed to store clothes and jewelry belonging to the deceased inside the sepulchre.

regions. This was the case of Taweret and Bes, household deities associated with conception and birth. According to various reports, Taweret was a mixture of crocodile and hippopotamus, with a female's breasts and a huge womb. In turn, Bes was a dwarf demigod with a huge head, often depicted wearing a mask.

Both magical and funerary texts mention a large variety of devils and other intriguing characters with grotesque shapes that tended to appear at night. One of the most feared was Apep, a giant serpent whose main enemy was the sun god, whom he ambushed as light diminished at twilight. Generally, he was unable to carry out his evil deeds as Set, watchful of any attacks from his solar boat, speared him with a javelin. Egyptians tended to attribute a red sky to the blood that flowed from Apep's injuries.

OSIRIS
The god of the dead, usually depicted in the form of a human wearing an Atef crown.

Official and private religion

In any case, religious worship among Ancient Egyptians was twofold: official worship, carried out in public, and private worship, practiced within the confines of individual households. Official religion focused on the worship celebrated at the temples. The Pharaoh was responsible for ensuring the benevolence of the gods to his people and the priests he appointed were responsible for caring for the temples. Gods lived in the images and sculptures there, and the Pharaoh maintained a permanent dialog with them. The relationship between the ruler and the gods was viewed as an exchange of equals. One inscription found at the burial site of Horemheb states: "The Pharaoh came to you [the deity], brought offerings and gave them to you so that you can give him all the lands or a similar gift." The exchange between the Pharaoh and the gods was regarded as a matrimony, involving love and mutual respect. Alongside official religion, private worship was practiced. Such religious activities were enshrined in a certain sense

The survival of the royal family was guaranteed. However, other humans had to seek survival in the afterlife by employing magical methods

of magic. This practice involved the use of amulets and figurines and particular adornments that protected their owners, in addition to the busts of ancestors who were venerated in households. In general, these private cults preserved ancient rituals from earlier religious eras.

The cult of the dead

Just like in other religions, the topic of death was a central feature of Egyptian religion. The kingdom of the dead was thought to be found in the lower regions of the world. As with other aspects of religion, various different versions of this topic exist depending on different regional beliefs and the preexistence of cults that preceded the region's inclusion in the Egyptian empire. The survival of the Pharaoh and his family after death was considered to be guaranteed. However, other humans from the lower levels of society had to seek survival in the afterlife by employing magical methods and by means of various rituals. Indeed, the journey to the afterlife, which was undertaken by boat, was full of hazards and dangers that had to be overcome.

For the most socially and economically advantaged sections of society, preparation for the afterlife occupied almost their entire existence and focused on the construction and furnishing of the ideal sepulchre. Reaching their desired destination involved identifying with the world of the gods, in particular Osiris; otherwise, Egyptians would not be allowed to board the boat that transported them to the afterlife, and they would spend eternity on earth, generally prowling their own tomb. According to funerary myths, the boat that transported the dead to the afterlife could carry hundreds of dead at a time; this was significant during times of war, for example, when mortality rates rocketed. Its size was said to be so great that its entirety could not be fully appreciated.

COFFINS
The coffin in which the embalmed body was placed was made from painted wood. This one was found in the necropolis at Thebes and is dated to 850 BCE.

Judgment of Osiris

Between death and inclusion into the afterlife, which some papyri refer to as the "journey through the perfect paths of the horizon," the deceased had to undergo a judgment phase, from which the Pharaoh was exempt. The deceased's heart was placed on a weighing scale. As a counterweight, Maat, the incarnation of integrity and order, was used. Maat was depicted in various ways: in hieroglyphs, as an ostrich feather and as the figure of a goddess—the goddess Maat—with a feather fastened to her head with a band. Thoth, the god of wisdom and justice, was responsible for presiding over the weighing process, in the presence of Osiris and 42 judges. When the scales balanced, the deceased was welcomed by Osiris. When a balance could not be reached, the deceased was given the opportunity to explain how he or she had behaved on earth.

In all representations of the judgment, there is a female monster known as "Devourer of the Dead" responsible for consuming those who failed to pass the test. The annihilation of the deceased was viewed as a second death, considered more serious than the first. In funerary texts, these deceased, who were in fact already deceased but who had died again, were sentenced to a range of punishments, as they symbolized a constant danger and threat to the living. While on earth, Egyptians did everything in their power to safeguard their earthly lives from this danger.

MAAT
Representation of truth and justice, Maat's head is adorned with the "feather of truth"; she served as a counterweight against the heart of the deceased during the Judgment of Osiris.

Burial paraphernalia and mummies

To mitigate these serious outcomes, burials were accompanied by a large range of material goods, ranging from food to jewelry. Throughout this strange journey, the soul of the dead would need food and, at the same time, offer food to the different deities with whom he or she would cross paths. As a result, the wealthy were more likely to be welcomed by Osiris

Between death and inclusion into the afterlife, the deceased had to undergo a judgment phase, from which the Pharaoh was exempt

than other mortals. Some wealthy Egyptians built large statues in their tombs for their soul to occupy while the judgment was in progress. Thus, they would not be forced to prowl their tomb until a favorable sentence was handed down. The mummy itself, wrapped with the utmost care and protected by a number of amulets, served to protect the deceased from any eventuality that could occur on their journey to the afterlife. Preserving the body, which was duly embalmed and interred, helped to preserve the soul in the afterlife. The mummy was placed in a coffin, made from wood; this comprised a box closed with a cover. In turn, the coffin was placed inside a sarcophagus, made from limestone, granite, basalt, or marble, depending on the social standing of the deceased.

Canopic jars were placed inside the burial chamber, close to the coffin. These funerary jars contained the viscera of the deceased, extracted during the mummification process. The jars were made from alabaster and were sealed with lids in the form of a human head. However, during the Ramesses Dynasty, they tended to reproduce the head of the four children of Horus: Imsety, protected by Isis, had a human head and protected the liver; Hapi, associated with Nephthys, had the head of a baboon and was responsible for preserving the lungs; Duamutef, protected by Neith, had the head of a jackal and was associated with the stomach; and Qebehsenuef, who represented Serket, had the head of a falcon and protected the intestines.

In all sepulchres, numerous figurines known as *ushabti* have been found. They were small statuettes capable of realizing all the effort necessary, including physical, to assist the deceased to follow the normal journey to the afterlife, into the arms of Osiris. The *ushabti* had to sacrifice their "life" to prevent the second death of the deceased, which was always associated with danger and suffering.

FUNERARY PARAPHERNALIA
Armlet featuring the sacred scarab, one of the jewels that accompanied the mummy of Tutankhamun in his tomb.

USHABTI BOX
Wooden box in which the figurines that served the deceased during his or her journey to the afterlife were placed. The image depicts the moment at which the deceased appears before Osiris.

The Creation Myth

The nature of Ancient Egyptian mythology varies, with versions depending on the different sources available and the different regions of the empire from which they came. However, evidence of rich and original myths have reached us, such as the creation of the world, represented in various wall paintings and carvings.

The origin of the world

Akin to all cosmology, the Egyptian variations gave comprehensive explanations of life. They maintained that the world had been created as a result of a conflict between the gods. Although no explanation is given for the cause of the argument, it was believed that Shu, god of the air, took Nut, goddess of the sky, in his arms, separating her from Geb, who was responsible for the earth. Thus, the separation of the upper and lower plains was established, while the boat of Osiris Re set sail through the body of Nut.

The journey of the sun

According to Egyptian belief, the journey of the god Osiris Re, incarnated in the solar disk, began at dawn. He then traveled the sky on his boat. He died at dusk and later sailed the underworld during the night.

VEGETATION
The leaves that cover the body of Geb, god of the earth, allude to the renovation of vegetation.

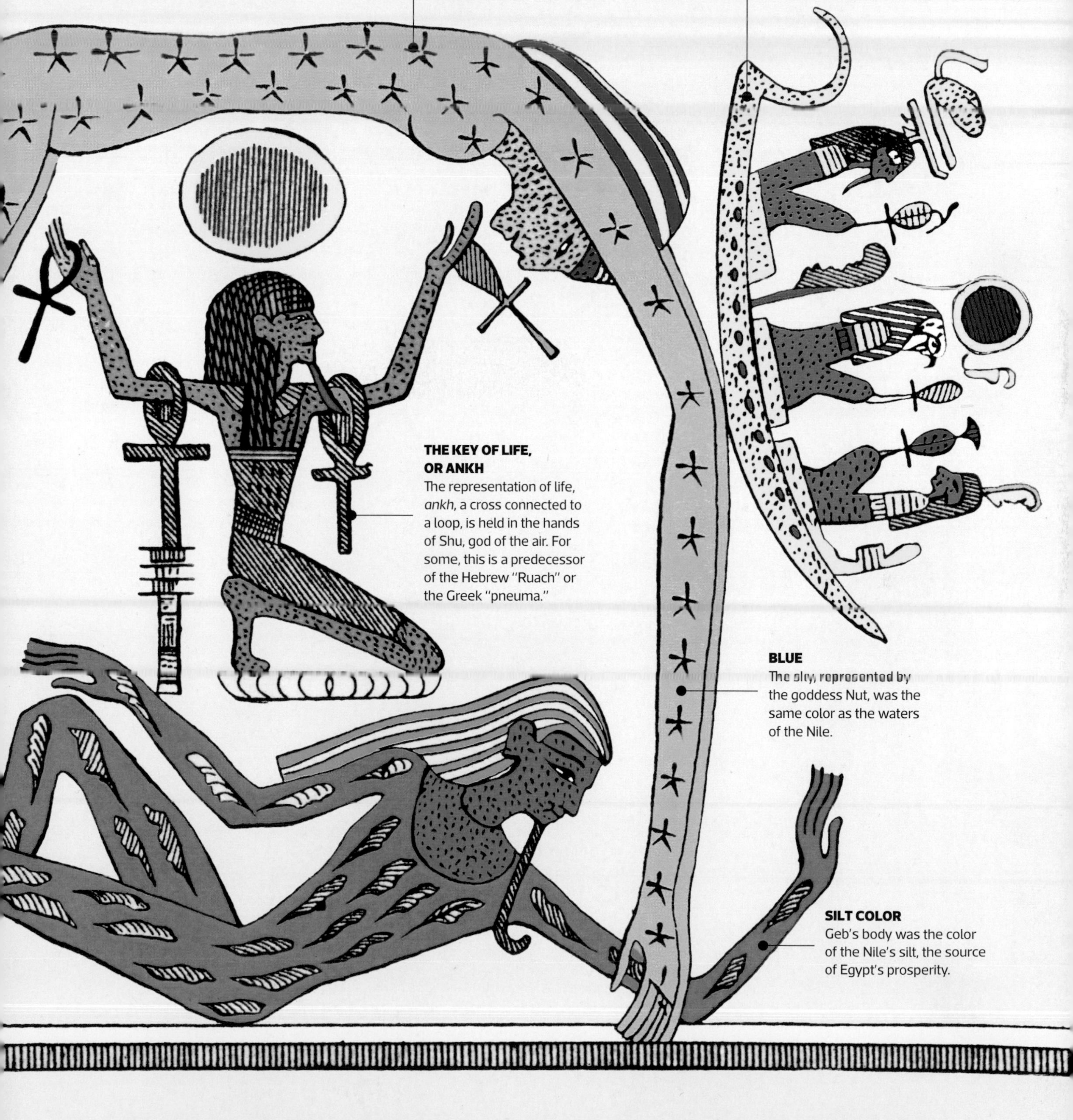

THE BOAT OF OSIRIS
Flanked by Isis and Horus, it navigates through the body of Nut, in a wall painting from the tenth century BCE.

THE STARS
Cover the heavenly body of the goddess Nut.

THE GOD RE
Aboard a boat, he travels the heavenly body of the goddess Nut.

THE KEY OF LIFE, OR ANKH
The representation of life, *ankh*, a cross connected to a loop, is held in the hands of Shu, god of the air. For some, this is a predecessor of the Hebrew "Ruach" or the Greek "pneuma."

BLUE
The sky, represented by the goddess Nut, was the same color as the waters of the Nile.

SILT COLOR
Geb's body was the color of the Nile's silt, the source of Egypt's prosperity.

Main Deities

The main deities were typically local, and corresponded to a particular region; in different regions the main deities tended to unite, forming triads. In numerous cases, the gods and goddesses broke geographical barriers and were worshipped throughout Egypt.

ATEF CROWN
A symbol of power, it was usually depicted in yellow and it was believed to assist the deceased in the process of rebirth.

WHIP
Symbol of authority and leadership, the Pharaohs used them in funerary and ceremonial rituals to identify with Osiris.

Osiris, Lord of Gods

God of resurrection and the fertility of the Nile, he presided over the council that judged the dead and determined their future. He was slain by Set, his brother. Isis recovered his body parts and brought him back to life. Horus, their son, was then born and avenged his father's death by banishing Set. The Pharaohs considered themselves descendants of Horus.

SOLAR DISK
As the daughter of Re, Isis is represented wearing a solar disk as a headdress flanked by two horns, attributes of the goddess Hathor.

The goddess Isis

Goddess of maternity and birth, her Egyptian name was Aset, meaning "throne"; the name Isis is of Greek origin. Her more common representation, in a sitting position with a solar disk in her headdress, can be traced back to the eighteenth Dynasty. Sister and wife to Osiris, the most important temple dedicated to her cult was on the island of Philae. She was also venerated at Giza as the Lady of the Pyramid.

THE DIVINE TRIAD
Osiris and Isis, and Horus, their son, comprise the divine triad at Abydos.

HATHOR
Identified as the mother god. Protector of women, music, love, and the deceased. She is depicted with the head of a cow.

THE CROOK
Employed by nomadic tribes to drive their animals, the crook was introduced as part of Egyptian iconography to represent (when shown with deities) guidance and protection of the people.

Amun, god of the universe

His name meaning "hidden," Amun was associated with the origin of the Universe. The main god of Thebes, he was united with the sun god Re, reaching the throne of the Egyptian pantheon during the New Kingdom. Numerous hymns and temples were dedicated to him, referring to him as the "one who makes himself into millions."

HORUS'S NECKLACE
Comprising various sheets of gold and colored glass, the divisions symbolize the different parts of a falcon's wing and plumage.

DEPICTION
He is often depicted as having a human form with a crown of long feathers. He can also be represented as having the head of a ram.

CANE
Like most other Egyptian gods, Horus holds a cane, or crook, in his hand as a symbol of guidance and protection.

Horus, god of unification

God of the sky, light, and kindness, he was the son of Isis and Osiris. Venerated throughout Egypt, he most commonly appears as a falcon or a man with the head of a falcon. He is also depicted as a sun with wings, and as a child with a finger covering his lips. A heavenly god and healer, Horus was respected as being responsible for the rise of the Egyptian civilization and was invoked during illness and legal disputes.

Local Gods

The main deities coexisted with a long list of local and regional deities. As the large states in the north and south emerged, many of them were associated with the more developed centers. Upon these foundations, a rich and varied pantheon was built.

Syncretism

Certain local deities were absorbed by other more powerful deities, creating compound gods in a complex process known as syncretism. Thus, various less known deities were worshipped in the same places as the gods that assumed their attributes, spreading their cult throughout the country.

Thoth, the god of wisdom

Venerated at Hermopolis, Thoth was the inventor of scripture and the sciences. He was also the god of the moon and ruled the path of the stars. He was depicted as having the head of an ibis or lunar disk, or even as a baboon.

BABOON
Thoth was shown as a baboon, a primate common to the semidesert or savanna.

Bast, protector of the home

Bast, or Bastet, was the goddess of household protection and as such symbolized harmony and happiness. She represented the warm rays of the sun and their beneficial effects. She was worshipped in Bubastis and in her honor, a large popular festival was held during which singing, dancing, and drunkenness were strongly featured to please and flatter the goddess.

MULTIFACETED GODDESS
Bast was depicted as a domestic cat, but when she was angered, she was depicted as a woman with the head of a lioness.

Apis, the bull

His main center of worship was Memphis, although he was also worshipped in Sais and Athribis. A symbol of virility and fertility, he occasionally replaced the traditional boat and transported the dead on his back on their journey to the afterlife.

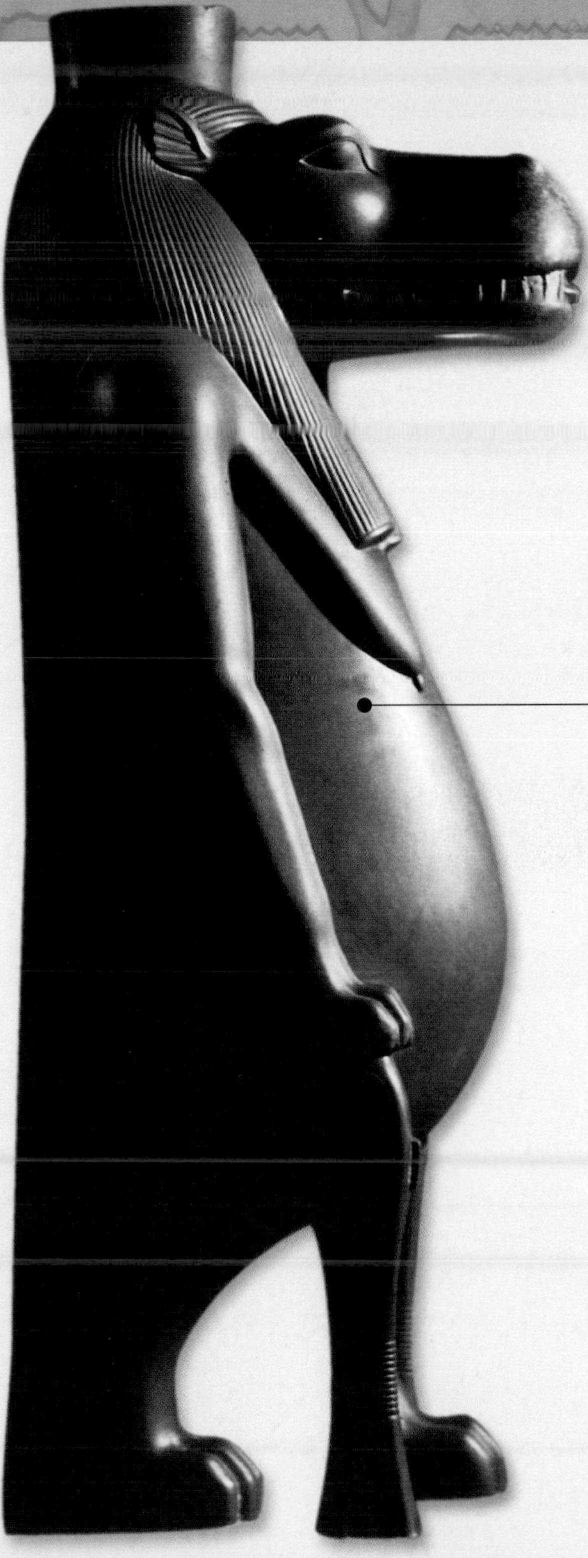

THREATENING
Bes, a dwarf, was depicted naked or barely covered by a loincloth, and with a long mane and shaggy beard.

Bes, a protective spirit

Bes was the tutelary (guardian) god of couples and could be found wherever women and children needed his care. His face, in which oriental features can be traced and which was linked to Mesopotamian-based cults, appeared during the New Kingdom as "the Lord of Punt" or "the Lord of Nubia." He has a threatening appearance, which he used to chase off evil spirits that attacked while the household's residents slept.

AMULETS
Amulets of Taweret, pregnant and with large breasts, tended to feature in the home to protect pregnant women and children.

Taweret, the great

Taweret, a cross between a hippo and a woman, with a lion's claws and the tail of a crocodile, was one of the most popular goddesses; her protective role was to oversee fertility and a successful pregnancy. Her grotesque appearance was probably attributable to an attempt to ward off evil spirits. Highly revered in Karnak, her cult extended to other territories.

HUMAN APPEARANCE
Ptah's most common representation is as a small man wearing a shroud, a cap on his head and a royal scepter.

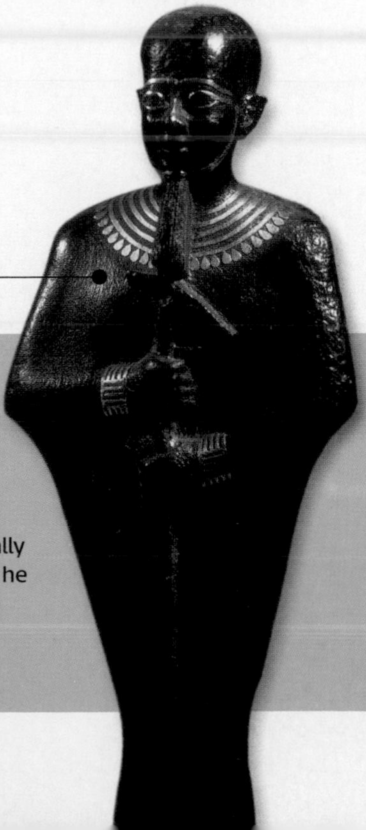

Ptah, the creator

Linked to carpentry and crafts, Ptah is associated with creating the universal order. He was also believed to have healing powers. He was especially worshipped in Memphis, where he formed part of the triad with the goddess Sekhmet and the child god Nefertem.

The Temple of Amun-Re

This temple, dedicated to the King of the Gods, was the largest and most famous of the four housed at the sacred complex in Karnak. Built during the Middle Kingdom, it was successively expanded and embellished by the Pharaohs of the New Kingdom.

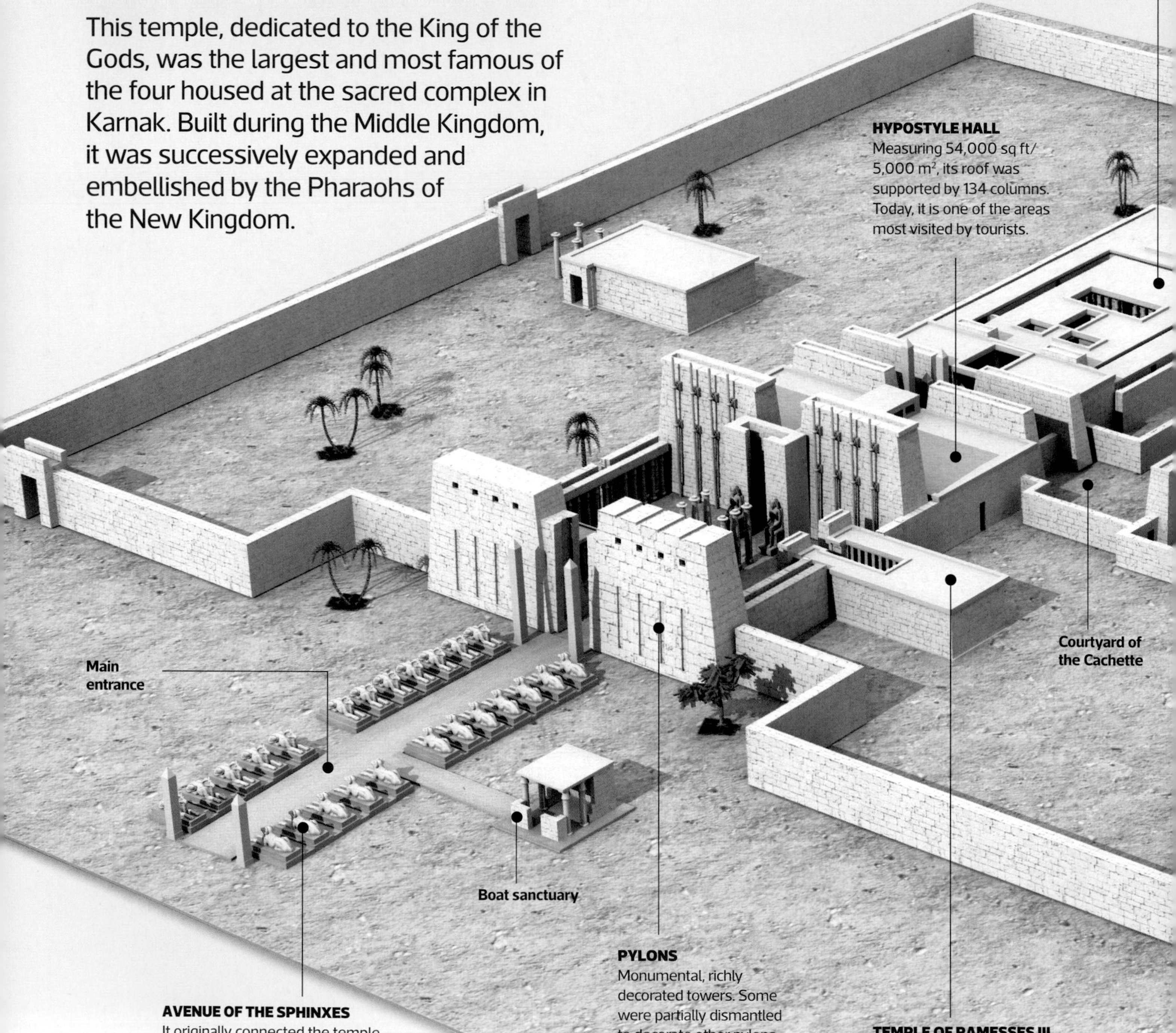

SANCTUARY
The heart of the complex, where the figure of the creator, Amun, was housed. At Egyptian temples, construction always began with the sanctuary.

HYPOSTYLE HALL
Measuring 54,000 sq ft/ 5,000 m², its roof was supported by 134 columns. Today, it is one of the areas most visited by tourists.

Courtyard of the Cachette

Main entrance

Boat sanctuary

PYLONS
Monumental, richly decorated towers. Some were partially dismantled to decorate other pylons.

AVENUE OF THE SPHINXES
It originally connected the temple with the pier on the River Nile; later, it linked the temple to the Temple of Amun in Luxor. It features 40 ram-headed sphinxes.

TEMPLE OF RAMESSES III
A small temple, built during the time of Ramesses III (twelfth century BCE).

King of Gods

It was Amenemhat I, founder of the seventeenth Dynasty, who united Amun, a deity worshipped in Thebes, with Re, the sun god worshipped in Heliopolis. Thus, Amun-Re was sworn the King of Gods, the most important deity of the second millennium.

HALL OF THUTMOSE III
A spacious room, the purpose of which remains unknown. The main hall housed the Karnak king list, featuring the chronological order of the 61 preceding Pharaohs, currently on display at the Louvre in Paris.

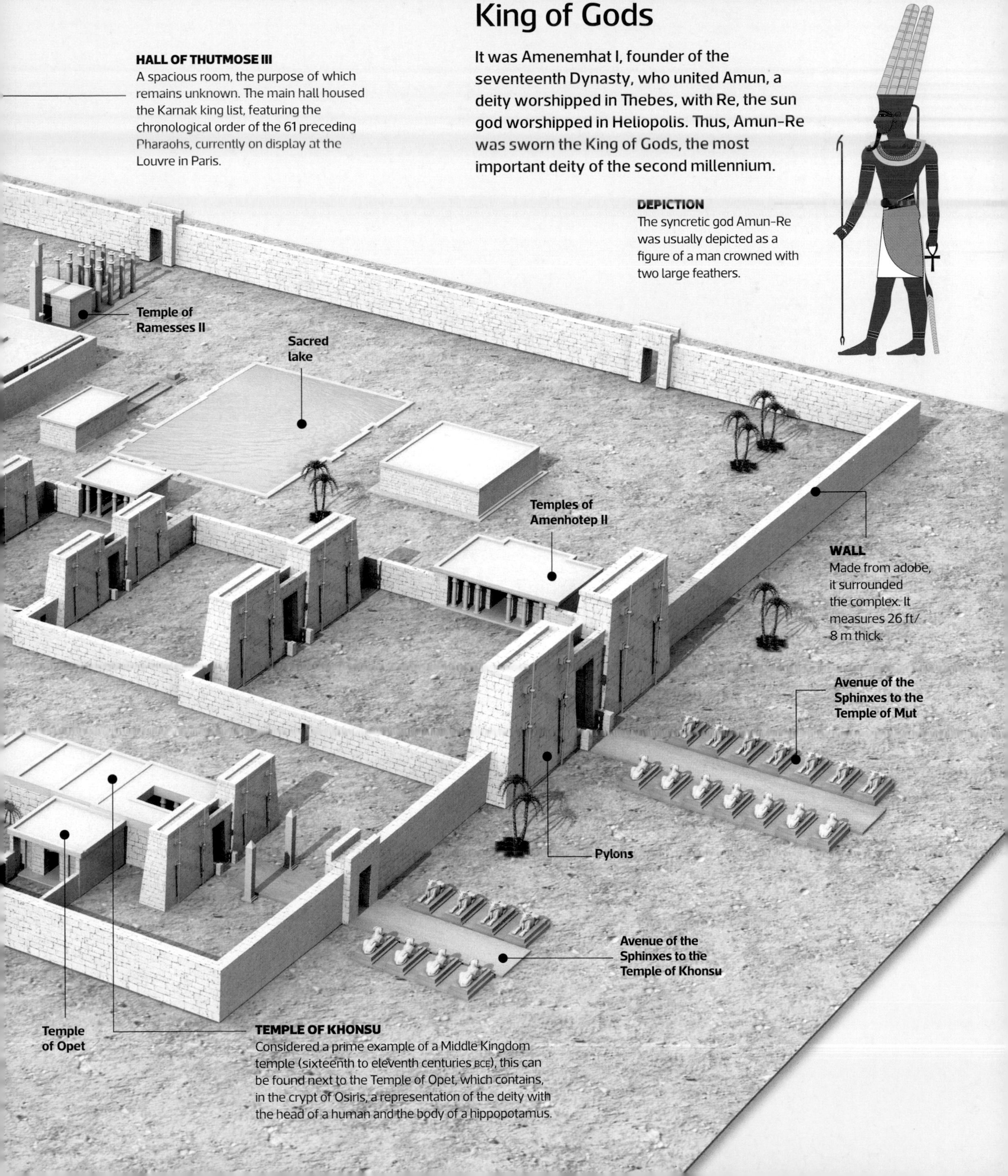

DEPICTION
The syncretic god Amun-Re was usually depicted as a figure of a man crowned with two large feathers.

WALL
Made from adobe, it surrounded the complex. It measures 26 ft/ 8 m thick.

TEMPLE OF KHONSU
Considered a prime example of a Middle Kingdom temple (sixteenth to eleventh centuries BCE), this can be found next to the Temple of Opet, which contains, in the crypt of Osiris, a representation of the deity with the head of a human and the body of a hippopotamus.

The Beautiful Feast of Opet

In Ancient Egypt, the main celebrations were religious in nature, with the Pharaoh playing an active role in them. One of the most important was the Beautiful Feast of Opet, which was held annually during the flooding of the River Nile. The festival celebrated fertility and regeneration, reaffirming the divine nature of the Pharaoh.

The procession

The celebration consisted of a procession that left the Temple of Amun at Karnak, and arrived at the same deity's temple in Thebes (Luxor), a journey of around 2 miles/3 km. During the reign of Hatshepsut (fifteenth century BCE), the procession passed through the Avenue of the Sphinxes, which directly linked both temples, and returned by boat on the Nile. During the reign of Ramesses II (thirteenth century BCE), both journeys took place on the river.

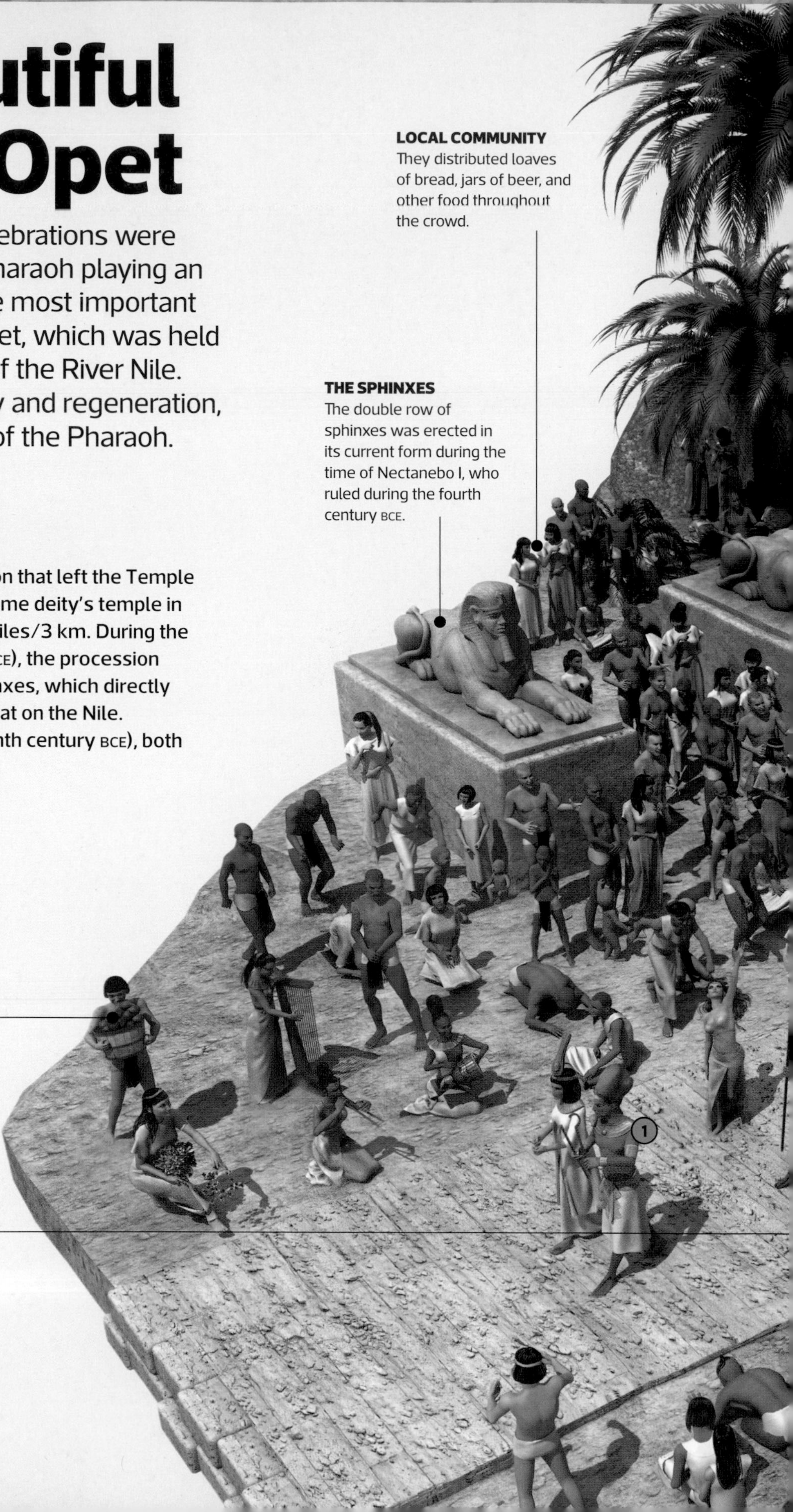

LOCAL COMMUNITY
They distributed loaves of bread, jars of beer, and other food throughout the crowd.

THE SPHINXES
The double row of sphinxes was erected in its current form during the time of Nectanebo I, who ruled during the fourth century BCE.

MUSICIANS
They played string, percussion, and wind instruments.

DANCERS
They danced around the procession to the sound of the music. They also performed acrobatics.

1

DURATION
The festival lasted for 12 days during the reign of Hatshepsut, increasing to 27 days under Ramesses II. The statue of Amun was left in Luxor for 24 days before returning to Karnak.

Temple of Luxor

The current arrangement of the Temple of Luxor was laid out by Amenhotep III (father of the famous Akhenaten), who ruled during the first half of the fourteenth century. Later, Ramesses II would expand it significantly. During the journey back to Karnak, the boat of Amun was paraded among the festival-goers through the impressive entrance to the temple.

1 The Pharaoh
Accompanied by his royal consort, he led the procession, through the jubilant throngs of people.

2 The priests
Covered in leopard skin, they walked in front of the boat of the god, chanting sacred hymns.

3 The boat of the god
The statue of Amun, within the ritual boat, was carried by priests wearing masks that represented the gods.

Giants. Two statues, 50 ft/15 m high and featuring the face of Ramesses II, guarded the entrance to the temple.

3

2

The Reforms of Akhenaten

Amenhotep IV changed his name to Akhenaten and unleashed a revolution that would change many aspects of the country. Although his objectives were merely religious, Akhenaten's reforms also resulted in a political breach and a radical overhaul of the taxes paid at the time.

A detested Pharaoh

Akhenaten built a new capital, Akhetaten (Tell el-Amarna), where he resided with the entire court, and erected a sacred temple to the cult of Aten. His revolution failed to please the clergy or the powerful families of Upper Egypt, and resulted in a number of internal conflicts. His policies were also resented. Following his death, all his reforms were abolished, the cult of Amun was restored and the capital moved to Memphis.

PORTRAITS
This monumental statue of Akhenaten shows the Pharaoh with very distinctive, almost caricature features: oblique, narrowed eyes, protruding lips, and a prominent chin, which are characteristics of the new Amarnian art.

New cult of Aten

Akhenaten broke with the previous ritual rule and formed a cult to a single god, the creator of the universe: Aten. He was depicted as a solar disk whose rays ended in the shape of human hands. According to this new ideology, the king merged with the sun to reappear each morning in his image. Open courtyards with altars were built as places of worship in order to greet the Pharaoh as the incarnation of the sun every day.

The royal family. Akhenaten, his wife Nefertiti, and his daughters worshipping the god of creation, Aten.

Amarna letters

Mainly written in Akkadian, the lingua franca and language of international diplomacy, these tablets contain correspondence between the Egyptian administration and its representatives in other regions. Thanks to them, it has been possible to establish what relationships existed between Egypt and Assyria, Mitanni, and the Hittites in Anatolia during the reigns of Amenhotep III, Akhenaten, Smenkhkare, and Tutankhamun.

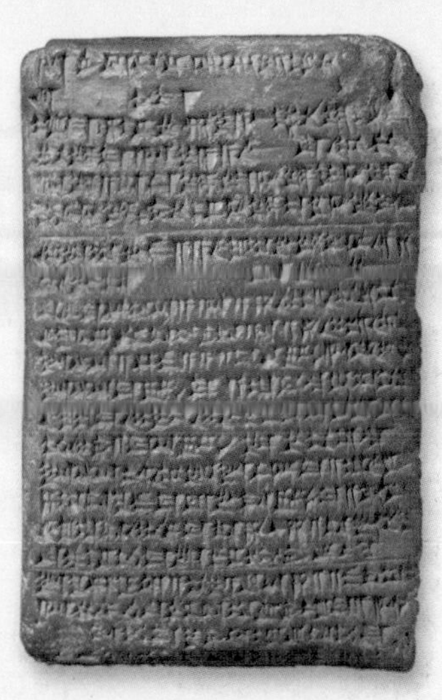

Aesthetic revolution

The artistic revolution encompassed all fields. In architecture, the new technique of the *talatat* was used; these smaller stone blocks could be transported by a single man, and made temple construction faster. The paintings and reliefs became more dynamic, and scenes of nature and the daily life of the court became predominant.

SUNK RELIEF
This new form of relief became commonplace, replacing bas-reliefs. Using this technique, figures were modeled inside the recess.

EMOTIONS
In this sculpture, the Pharaoh can be seen kissing his daughter. It is an example of how new topics were broached, such as the private life of the Pharaoh, which were previously unheard of.

Journey to the Afterlife

In Ancient Egypt, in order to successfully reach the afterlife, the "spirit" of the deceased person, the *ka*, had to pass the Judgment of Osiris. The Book of the Dead was used to help with this journey, and comprised a collection of spells and magic formulas that served this purpose.

"NEGATIVE CONFESSIONS"
Any statements of innocence alleged by the deceased's *ka* while defending their passage to the afterlife received this name. These beliefs demonstrate that death was also seen as a moral event.

The Book of the Dead

Dates back to the New Kingdom, although its origins have been traced to the Old Kingdom. Its content sets out the journey that the spirit of the deceased must undergo to ensure immortality. A copy of this compilation—no canonic original exists, as its length varied according to the wealth of the deceased—was placed in the sarcophagus or the burial chamber of the deceased.

OSIRIS
This god presided over the council that judged the deceased's spirit. By passing the judgment, the deceased was introduced into the light of divinity.

SCALE
The scale decided whether the *ka* passed the Judgment of Osiris. On one of the scales, the deceased's heart was placed; on the other, the feather of Maat, goddess of truth.

KA
The spirit or vital energy of the deceased—his or her *ka*—was taken by boat over the Nile until it merged with the *ka* of Osiris.

Death as the dawn

Ancient Egyptians knew the Book of the Dead under various guises, such as "Book of Coming Forth by Day" or "Book of Coming Forth by Light." These names were inspired by the belief that, for the *ka*, death represented a dawn, symbolically watched over by the reappearance of the solar disk in the sky.

The scarab beetle, an object of worship, held the solar disk with its pincers.

CHAPTER 125
The most important chapter in the Book of the Dead, given that it describes the judgment ritual to the deceased. Thoth, the god of scribes, records the verdict.

Detail of a papyrus from the Book of the Dead.

Burials

An Egyptian funeral comprised various basic stages. The first was the wake at the house of the deceased, followed by the procession of the coffin to the necropolis (across the Nile), a ceremony before the tomb, and, finally, interring the mummified corpse.

Expectation

The display of the funerary paraphernalia of the most famous deceased aroused great curiosity during the procession of the coffin.

CANOPIC CHEST
In which the viscera of the Pharaoh were transported.

AY, THE PRIEST
He led the funerary procession of Tutankhamun and acted as chief priest.

ANKHESENAMUN
Tutankhamun's widow laid flowers on his mummy; these were still visible when Carter opened the sarcophagus.

The funeral of Tutankhamun

As can be seen in this illustration, the funeral procession at the burial of the young king was led by Ay, his adviser and successor. He was followed by members of the royal family, army generals, and high dignitaries from the court. They were accompanied by the mourners. The procession traced the path from the Temple of Thebes to the Valley of the Kings, on the opposite side of the Nile. The exact location of the tomb was a secret, unknown to the procession.

THE SARCOPHAGUS
The mummy was transported inside the sarcophagus. Once at the tomb, it was removed to perform the last rituals before being deposited in the mortuary chamber.

MOURNERS
This group of women followed the procession in tears, lamenting the death of the deceased.

SERVANTS AND SLAVES
Responsible for carrying the funerary paraphernalia, they were also responsible for transporting the sarcophagus and canopic chest.

Mummies

To ensure the journey to the afterlife was a success, it was essential that the integrity of the corpse was preserved to the maximum. The embalming or mummification process was a part of this belief.

The *ushabti* were small funerary statuettes included in the paraphernalia.

Funerary paraphernalia

The deceased was interred with numerous objects for preserving the level of comfort enjoyed on earth. The paraphernalia included food, clothing, furniture, articles of personal hygiene, ritual objects (such as figurines of protective deities) and other items related to the deceased's interests.

To help the deceased's journey, they usually included a pair of sandals and a miniature solar boat.

Embalming process

Initially, only the most powerful people were mummified, as significant costs were involved, but toward the seventh century, the process became more widespread. Mummification lasted 70 days, during which time different religious ceremonies were held.

NATRON

The corpse was filled with a type of natural salt—natron—to dry it out. The body was then washed and oils were applied to make the skin flexible.

LINEN BANDAGES

To dress the body, strips of linen impregnated with a resinous liquid were used; the liquid served to stick the strips together and harden them as it dried out.

Mummy of Ramesses II, found in the complex at Giza.

Inside the sarcophagus

Once mummified, the deceased's body was placed inside a coffin and the bandaged face was covered with a mask that represented him/her. If the deceased had been wealthy in life, the mummy was placed inside a number of successive coffins, one inside the next. In the case of Tutankhamun, for example, three coffins were found inside the sarcophagus.

Preservation of organs and viscera

The organs extracted from the corpse during the mummification process were washed and covered in natron and placed inside canopic jars. These jars represented the four children of Horus, and were buried with the deceased. From left to right: Falcon, protected the intestines; Jackal, the stomach; Baboon, the lungs; and Man, the liver.

HOOK
A hook was introduced through the throat or nose to remove the brain. The viscera, with the exception of the heart, were removed from the side after an incision was made using a stone knife.

OUTER COFFIN
Reproduced the face of the deceased and was decorated with painted images and texts.

FUNERARY MASK
A golden or painted mask reproduced the idealized features of the deceased and was positioned above his or her face.

PECTORAL
A wide necklace was placed over the deceased that represented the image of Osiris.

INNER COFFIN
Once bandaged, the body was placed in the first coffin. An image of a winged scarab beetle was placed on the chest.

AMULETS
Jewelry and emblems belonging to the deceased, in addition to other objects, were placed inside the mummy's coffin in order to protect the person inside. Below, two examples of amulets.

SARCOPHAGUS
Coffins were placed in a stone sarcophagus, sealed with an equally heavy slab, which was then placed inside the funerary chamber.

The Mummification Ritual

The embalming process was long, complex, and delicate, and all its components were subject to a strict series of rituals. Different techniques were used, depending on the circumstances of the deceased and his or her economic power.

THE LECTOR-PRIEST
Dressed as Anubis, the lector-priest ("controller of mysteries") recited the instructions and magic formulas for each step of the process.

Main steps

1. First, all the viscera were removed from the body, as they decay fastest.

2. Next, the corpse was filled with natron, a mineral that removes moisture. It was left to work for 40 days, so that all liquid would be removed to fully dry out the body.

3. The mummy was washed, filled with a mixture of resin and linen, and anointed with ointments. Next, it was wrapped in several layers of linen.

4. Finally, the mummy was placed inside a coffin (or in several coffins, one inside another), the face was covered with a mask, and the body was adorned with jewelry.

REBIRTH
The goddess Hathor receives Tutankhamun after his death, reviving him by placing a sacred cross (*ankh* or "key of life") close to his nose. Anubis, the god of death, can be seen just behind the Pharaoh.

In secret

Ancient Egyptian sources did not reveal mummification details, the techniques for which were highly secret and were only discovered through Herodotus.

INCENSE
Incense was burned to perfume the air, while the task was carried out as quickly as possible.

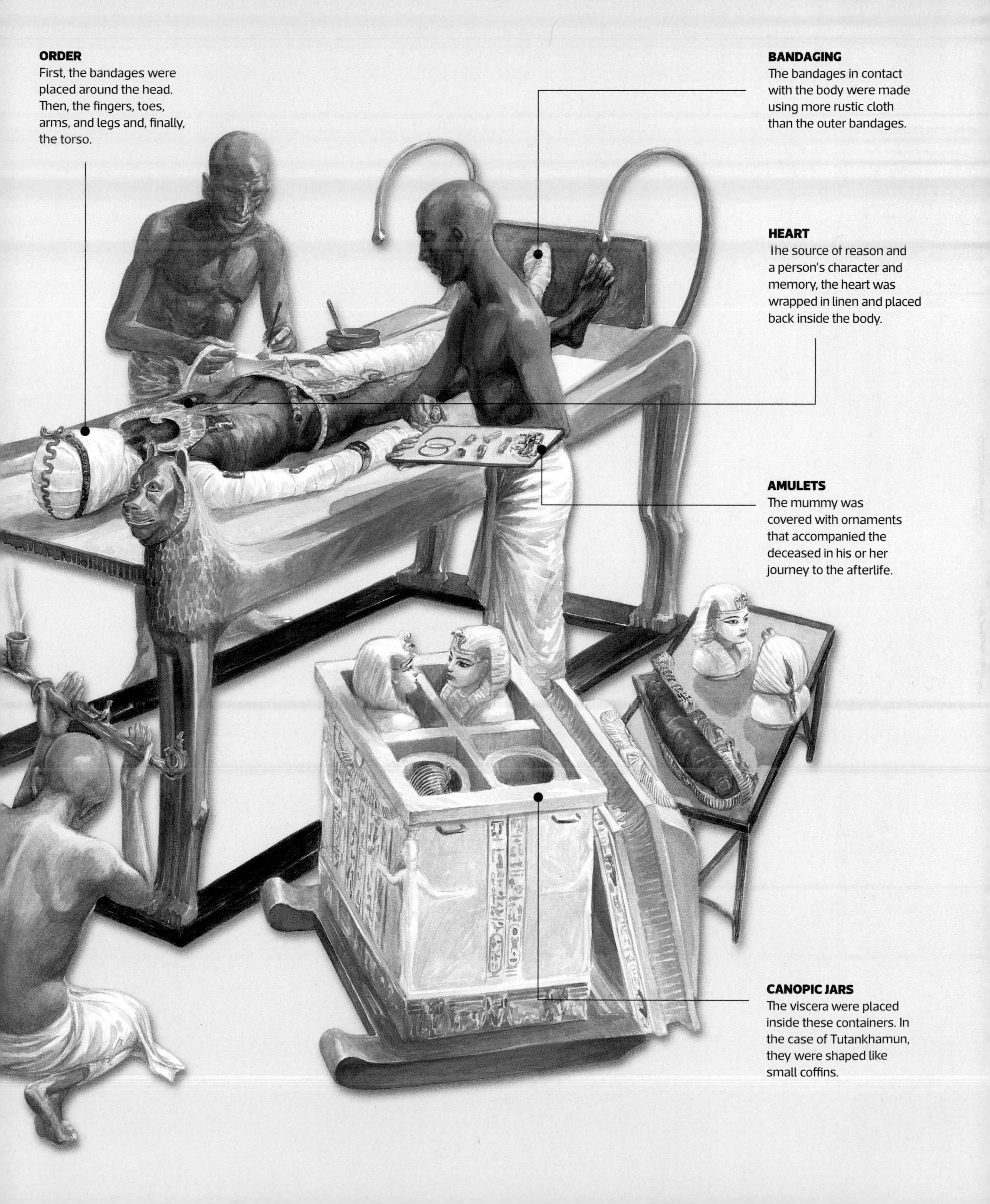

ORDER
First, the bandages were placed around the head. Then, the fingers, toes, arms, and legs and, finally, the torso.

BANDAGING
The bandages in contact with the body were made using more rustic cloth than the outer bandages.

HEART
The source of reason and a person's character and memory, the heart was wrapped in linen and placed back inside the body.

AMULETS
The mummy was covered with ornaments that accompanied the deceased in his or her journey to the afterlife.

CANOPIC JARS
The viscera were placed inside these containers. In the case of Tutankhamun, they were shaped like small coffins.

Private Tombs

The Egyptians had three types of tomb: pyramids, mastabas, and hypogeums. Hypogeums were underground tombs, carved out of the rock, which were common in the New Kingdom. They were used by members of the nobility as well as the Pharaohs themselves.

The Tomb of the Vineyard

This is how the hypogeum of Sennefer (TT 96), mayor of Thebes, came to be known. Buried here in the fifteenth century BCE, it was explored for the first time in 1826.

HIDDEN CHAMBER
Facing the direction in which the deceased should follow in his or her journey to the afterlife (east to west), it was the only space family members could access after the burial.

ANNEX
With a small 11-sq ft/1-m² pillar at the center, it is believed this space was used to store offerings.

CORRIDOR
An antechamber with a mortuary chamber. It measured 6 ft 6 in/2 m wide, 33 ft/10 m long and 13 ft/4 m high and was decorated with daily scenes featuring Sennefer.

Antechamber

Valley of the Nobles

The greatest concentration of private tombs in Thebes can be found in the Valley of the Nobles, in the necropolis of Sheikh Abd el-Qurna. These tombs are cataloged by number, preceded by the acronym TT (Theban Tomb).

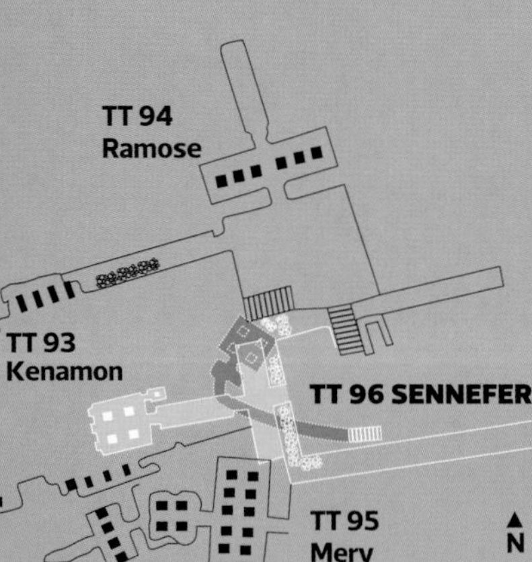

Hidden chamber
Mortuary chamber

The antechamber

Its ceiling was completely covered with painted bunches of grapes. Visitors would get the sense of being in a cool vineyard when below this.

Mortuary chamber

The irregular shaped roof was held in place by four pillars carved from the rock. Among the decorative paintings, the one depicting a vine and bunches of grapes is particularly noteworthy.

24 ft 6 in/7.5 m

6 ft 6 in/ 2 m

COLUMNS
They featured the image of Sennefer with an amulet in the shape of a double heart, as a sign of his loyalty to the Pharaoh.

Access to the hidden chamber

Lobby

The Garden of Amun

STAIRCASE
Carved into the sandstone rock, it comprises 44 steps. It goes 40 ft/12 m underground.

CORRIDOR
It was very low, measuring just 4 ft 5 in/1.35 m high and 4 ft/1.2 m wide. Its height forced anyone entering to bow reverently.

The Garden of Amun

This painting, located in the lobby, reaffirms the title of which Sennefer was most proud: the supervisor of the gardens of the god Amun.

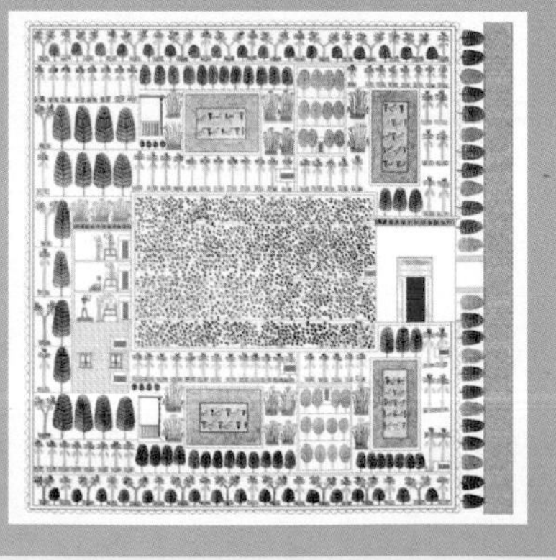

Sarcophagi

The treatment that the deceased's body underwent in preparation for the passage to the afterlife included being placed in a sarcophagus that, like other artistic elements, would evolve throughout the history of Ancient Egypt to adapt to the trends and requirements of each period.

Sarcophagi and coffins

The sarcophagus was a central feature of the funerary chamber, the space designed to house the corpse of the deceased. From the third Dynasty, from when the first known sarcophagi date, until the fifth Dynasty, they also served as coffins, and directly contained the body of the deceased. Later, the sarcophagus was built to house a coffin that contained the mummy.

Coffins

Wood was most commonly used to make coffins, although some were also made from reeds, animal skins, and precious metals. Just like the sarcophagi, they evolved, and royal coffins moved away from austere appearances and rectangular shapes, to anthropomorphic shapes featuring bright decorations.

Coffin of Henuttawy. From the twenty-first Dynasty. Anthropomorphic in shape, its entire surface was painted.

Sarcophagus of Ramesses IV

During the Ramesses Dynasty, sarcophagi underwent another transformation. They returned to a cartouche shape with a slightly curved cover. This is the case of the monumental sarcophagus of Ramesses IV, made from red granite. The Pharaoh's tomb is located in the Valley of the Kings.

Box for canopic jars. From the First Intermediate period, to prevent their theft when the chamber was looted, the boxes for canopic jars were inserted inside the sarcophagus, along with the coffin.

RELIEFS ON OUTER WALLS
The outer reliefs show images from the Book of Gates, a sacred text throughout the New Kingdom that narrates the journey of the deceased to the afterlife.

THE OUTSIDE
It measures 11 ft 6 in/3.5 m in length. The cover is adorned with human representations of sacred snakes that flank the king, incarnated as Osiris.

THE INSIDE
On the inside of the sarcophagus, scenes from the Book of the Earth are depicted that narrate, among other things, the raising of the solar disk from the depths of the earth.

Evolution

From the first sarcophagus designs found, belonging to the third Dynasty, the shape, materials, and decorations on the walls gradually evolved.

Old Kingdom
Sarcophagi were invented for the Pharaohs. They were rectangular with a flat cover. If there was any decoration on the outside, it portrayed the palace's facade.

First Intermediate Period
Two important innovations were introduced: the incorporation of the box for the canopic jars, and the sarcophagi were much bigger to deter looting.

Middle Kingdom
Built using single slabs of stone. The sarcophagi of the queens of Mentuhotep II are worth particular mention, as they were decorated on the inside and outside with carvings of funerary scenes.

New Kingdom
A new oval shape was adopted for the part containing the head. The outside was decorated with images of goddesses on the short sides or corners, and gods on the long sides.

Sarcophagus of Thutmose IV. Large in size, it could house several coffins. In the image, protector gods accompany him in the passage to the afterlife.

Mummified Animals

The art of Egyptian mummification was not limited to humans; many animals were also mummified. The purpose of embalming was not always the same: pets, select cuts of meat, and sacred animals were mummified. Mummies were also given as offerings.

Cult of Apis

When priests found an Apis bull, it was transferred to stables close to the Temple of Ptah where it lived with a harem of cows. The faithful would visit the animal in search of answers and, when it died, a 60-day period of mourning began.

Relationship with animals

The relationship between Egyptian society and animals was special. In the Egyptian approach to the world, animals were of equal worth with humans, and many gods were represented with the body of a human and the head of an animal: mammals (the cow, Hathor), insects (the scarab beetle, Khepri), birds (the dove, Isis) and many more. Therefore, it is hardly surprising that animals were also mummified.

MEATS

As funerary offerings, select parts of the animal were given to the deceased to provide them with sustenance in the afterlife. The leg of the cow, one of the most valued parts, was mummified so it lasted as long as the mummy of the deceased itself.

Food for the afterlife. Food offerings in a wall painting.

Sacred animals

Some animals were seen as a manifestation of the essence of a god. In such cases, the chosen animal lived at the temple and was cared for until its death. It was given offerings and, when it died, its body was embalmed.

Pets

Dogs, cats, vervet monkeys, baboons, and gazelles were all common pets in Ancient Egypt. Their mummies were buried with their owners so that they could enjoy their company in the afterlife. They were often even buried in their own coffin.

Embalming

There were various different embalming methods, the quality of which also varied. The most sophisticated technique involved gutting the body through an incision and drying it out using natron. The animal's body was then filled with linen cloth.

Dog's head. The god Anubis, associated with

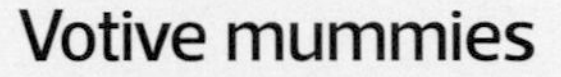

embalmers, was often depicted as a dog.

Mummified monkey. The embalmed animal's body was usually wrapped in linen bandages.

Votive mummies

Many species of animals were considered sacred: bulls, rams, dogs, ibis, falcons, and cats, to mention just a few. Believers could present a deity with an offering in the form of a mummified animal. This practice was limited almost exclusively to the Greco-Roman period.

Mummified cat. Incarnation of the goddess Bast. Many cats were mummified during the Greco-Roman period.

MUMMIFICATION
Occasionally, the body of the mummified animal was wrapped in papyrus and reeds.

ORACLE
The sacred animal was seen as an oracle that responded either directly or indirectly to the problems of the faithful.

Sobek. God of water and fertility, he was represented as a crocodile.

Fertility Statues

The large number of statues of naked women found in Egypt demonstrates the importance of female fertility. These images, used in domestic cults, guaranteed conception and protection of the mother and child during childbirth.

Common features

Generally, the figures were small and made from materials like clay, faience, wood, or stone. Normally, they featured a necklace and had a very prominent pubic triangle.

Influence in the afterlife

Their influence in the world of the living aside, some of these figurines have appeared in funerary monuments. Although initially they were identified as concubines at the service of the deceased, this interpretation was disproved when they were also found in the tombs of women and children. It is believed that they served to assist the dead during rebirth in the afterlife.

FACELESS
Accurately representing the face was considered least important of all the body parts.

CLAY MODEL
Clay statuette from the Second Intermediate period, around 1600 BCE.

PREDYNASTIC PERIOD
Figurines have been found dating back to the Predynastic period. Other ancient civilizations produced statuettes that served the same purpose.

RUGGED
The lines are rugged and the female sexual attributes are clearly distinguishable.

Help required

As many women died during childbirth, it is not surprising that they sought assistance from fertility figurines. They were also used as votive offerings, especially at temples dedicated to Hathor, the goddess of fertility, sexuality, and birth.

Taweret. Vase featuring the image of Taweret, a deity connected with birth. She was usually represented as a cross between a hippo and a woman.

HAIRSTYLE
Some examples of fertility figures show off very ornate hairstyles.

SVELTE
Woodwork made it possible to make the figures more svelte than those chiseled from stone.

NUBIAN WOMAN
Naked female suckling her child. She is accompanied by three monkeys, one on each shoulder and another at her feet.

HIGH RELIEF
Unlike the others, this statuette was worked as a high relief. She appears more authoritarian and stricter than the others.

CARVED FROM WOOD
Statuettes carved from wood atop a pedestal crafted from the same material are typical of the Middle Kingdom.

Amulets

Ancient Egyptians surrounded themselves with protective amulets in their daily lives and during their journey to the afterlife. The majority of these objects that have survived until now were discovered in funerary monuments.

Scarabs

The scarab or dung beetle was considered a symbol of the sun god, an incarnation of life that rose from the underworld each day. It was converted into the most popular amulet and seal, with countless examples made from the widest range of materials imaginable.

Symbolism of the materials

The representation in the amulet was just as important as the prime materials used to produce it. They were made from a wide variety of materials including metals, different types of minerals and precious or semiprecious stones, and man-made materials.

Wadyet eye

The Eye of Horus, hurt by his rival Set and cured by Thoth, is a symbol of plenitude and health. It became popular as an amulet given its ability to ward off darkness and provide protection against evil.

Tyet

Symbol known as the "the knot of Isis." Often made from red stone, it was used as a funerary amulet and was hung from the neck of the deceased. The *tyet* was associated with concepts of resurrection and eternal life.

PRESSED AGAINST THE BODY
Amulets such as this pectoral were worn pressed against the body, often hung from the neck, so that their power would be effective.

Uraei goddesses

These two upright cobras found in the tomb of Tutankhamun represent Sepulchre and Nekhbet, goddesses of Upper and Lower Egypt. With their poisonous breath, they warded off evil and were also a symbol of the Pharaoh's power.

Djed

Prehistoric fetish in the form of a totem. It became a symbol of stability and durability; as such, it was highly regarded as an amulet. During the New Kingdom, it was associated with Osiris and was said to represent his backbone.

Heart amulet

In the Egyptian world, the heart was the source of reason, willpower, and an individual's personality; it was left inside the deceased's body during the mummification process. A heart shaped amulet was placed on the body to silence the evil deeds an individual may have done in his or her lifetime.

COMBINATION OF AMULETS
By combining several amulets, their magic purpose could be changed, either increasing or modifying it. Especially during the Late Period, embalmers placed a significant number of amulets between the mummy's bandages to ensure the deceased would be reborn.

Influence on Christian Iconography

Several features of Egyptian religion formed part of a legacy for subsequent civilizations. The influence of symbols and Egyptian myths on the creation of Christianity is clear. Certain icons were replicated and continue to form part of present-day Christianity.

Egyptian Christianity

The origins of the Coptic religion can be traced to Ancient Egypt. From a very early age, the distinctive and unique features of its liturgy, art, symbolism, and architecture were influenced by their surroundings. Furthermore, the Coptic language and script, in which all its literature is published, can be traced back to the demotic language.

St. George and the dragon

In Late Period representations of the clash between Harpocrates (the child god Horus) and Set, Horus appears as a Roman soldier on horseback, and Set is depicted as a crocodile. This image may have given rise to the later representation in which St George, on horseback, slays the dragon with his spear.

Judgment of Osiris. A scene from the Book of the Dead.

The mother of God

A sculpture by Michelangelo representing the Mother of God with Jesus as a child. This representation, characteristic of Christian tradition, was preceded by Egyptian images of Isis feeding Horus as a child.

The Virgin and Jesus as a child. Marble sculpture by Michelangelo dating to 1501.

Isis and Horus. Statue from the eighteenth Dynasty.

The key of life

The *ankh* was adopted by the Copts and was associated with Christian symbolism. It is a hieroglyphic symbol that represents life, both in ordinary writing and in religious and symbolic writing and representations.

"Ankh" mirror. Piece from the tomb of Tutankhamun.

Coptic funerary stele. Bas-relief depicting various Coptic crosses and a boat.

Psychostasy (or weighing of souls)

To the left, a scene from the Egyptian Book of the Dead depicts Anubis overseeing the scales during the Judgment of Osiris. Sobek, a god with the head of a crocodile, is prepared to devour the souls of the condemned. In Christian tradition, represented in the image to the right, God is the one to pass judgment, Anubis is replaced by the archangel Michael, and Sobek by the devil, who is always prepared to take the sinner to hell.

Judgment. Painting from the thirteenth century CE, from Valle de Ribes in Spain.

5

CHAPTER

ART AND CULTURE

CULTURAL LEGACY

Ancient Egypt's artistic and cultural wealth is reflected in the pyramids and temples that aroused admiration among contemporaries, as well as in beautiful paintings, reliefs, sculptures, and literary works.

It was the Greeks, starting with the famous historian Herodotus, who were amazed when they discovered Egypt and who sought to assimilate its cultural legacy. They were responsible for communicating what they found to the world. Among the pre-Socratics, several Upper Egyptian cults left their footprint and were transformed into religious and philosophical principles that were later adopted into Classical Greek thought and proliferated throughout the West. Although the Greeks discovered the cultural wealth in the land of the Nile, it was the Romans who made it fashionable. After conquering Egypt in 30 BCE, Rome was swamped by Egyptian artifacts, considered attractive because of their "exoticism." Many Roman sculptors dedicated their lives to modeling Egyptian lions and sphinxes. Around the mid-fourth century, Rome boasted several obelisks and two pyramids.

Also, the Jews, whose presence in Egypt as slaves forms part of the first texts of the Old Testament, assumed many elements of the Nile culture. Taking a look at the writings of Sigmund Freud about the figure of Moses is enough to understand the deep bond between the cult of the Pharaoh and the first monotheistic religion. Christianity, founded among the Jewish communities in Ancient Palestine, also inherited important features of Egyptian culture. And in their westward march toward the Iberian Peninsula, Muslims also settled around the banks of the Nile, assuming the Egyptian legacy, integrating and transmitting it to other cultures.

Colossal constructions

Pyramids, the quintessential symbol of Ancient Egypt, were erected as an eternal testament to this yet-to-be-discovered legacy. A theocratic,

PHARAONIC PORTRAIT Ramesses I (nineteenth Dynasty) portrayed in a wall painting at his tomb in the Valley of the Kings.

perfectly organized empire, the social and political structure of which was "pyramidal," is reflected in the profile of these monuments, the construction of which still defies modern engineering. Their colossal size is a testament to the scale of the kings' wealth and power; the monarchy was capable of subjugating thousands and thousands of people to work year after year, extracting blocks from quarries and dragging them to construction sites, placing them on top of one another with an unparalleled level of artistry, until the tomb was ready to receive the mortal remains of the Pharaoh. The monarch was considered a divine being who, through being the master and owner of not just the lives and properties of all his subjects, but also the holder of the laws and secrets of the cosmos, abandoned earth to return to the land of the gods, from where he had descended.

To many anthropologists, the association of the Pharaoh with a supreme, divine being, incarnated in a human body, underlines the origins of monotheism and the Christian myths of rebirth and resurrection. Similarly, the judgment to which the souls of the dead were subjected on their journey to the afterlife has left its mark on all monotheist eschatological visions, which align with the concept of a final judgment.

A different perspective

The peak of the pyramid in which the Pharaoh was interred represented the final point in his soul's journey to the heavens. As the pyramids were built while the Pharaoh was alive, their peak also reminded the world of the living that the monarch ruled the destiny of all living beings from above. The mummification process itself, designed to make the ruler's

PYRAMIDS
The great pyramids of Giza are the most emblematic symbol of the social and religious sensibility of Pharaonic Egypt.

The paintings and drawings that decorated the pyramids are the epitome of simplicity; they also transmit an undeniable sense of solemnity

body eternal, symbolized his continuity in the afterlife. As Egyptians believed that preserving a body did not suffice, sculptors were tasked with carving a likeness of the monarch in everlasting granite. Interestingly, the word used to define a sculptor's job meant "he who preserves life." Wall painters, ceramic artists, and goldsmiths attested this calling to eternity with their crafts. The belief that the sovereign was as important as the gods, in addition to the belief that mortals were insignificant at his side, led Ancient Egyptian artists to create colossal reproductions to replicate such grandeur.

In paintings, which depict extraordinary likenesses, the reproduction of the Pharaoh in images was larger in size than other people. In Europe, during the High Middle Ages through to the Renaissance, religious Christian paintings also drew upon this convention. To represent their grandeur, Christ, the Virgin, and the saints were depicted as being larger in size than other beings. In both instances, this was not attributable to the artists ignoring the technique of perspective, but that "their" perspective of space, and of time, responded to a vision of the world as completely stratified and hierarchical.

Paintings and drawings

The paintings and drawings that decorated the pyramids are among the most beautiful works of Egyptian art. While they are the epitome of simplicity, they also convey an undeniable sense of solemnity. Artists, convinced that adding greater beauty and authority to their subject was useless, given the superior nature thereof, sought to utilize a naturalistic realism or geometric symbolism. Equally, the recreation of nature

WALL PAINTINGS
The guardian of the fifth gate of the Domain of Osiris, in a wall painting found at the tomb of Nefertari (nineteenth Dynasty), in the Valley of the Queens.

and the proportion of the composition demonstrated perfect balance, transmitting both life and eternity at the same time.

This combination of geometric regularity and a keen eye for replicating nature are characteristic of all Egyptian art. Where this can be best seen is in the reliefs and paintings found on the walls inside the sepulchres. It is worth noting that this art was not created to be "pleasing," but to fulfill a predominantly religious purpose: the works of art, saved for all eternity at the heart of the pyramids, were never considered for viewing by humans, but by the soul of the deceased and the gods.

COLOSSI OF MEMNON
Dated to the fifteenth century BCE, this gigantic figure is part of a colossal series of sculptures similar in size to those found at the Luxor Temple.

Singular realism

The realistic nature of the reliefs and wall paintings provides an extraordinarily illustrative reflection of just what life was like in Egypt thousands of years ago. However, Egyptian artists had a completely unique way of depicting real life. They did not seek beauty, but perfection, in line with their desire to render their subjects eternal. They drew from memory and in compliance with strict rules, which ensured perfect clarity of all the elements in the work. In reality, their methods were more akin to those of a cartographer than a painter. This affected their depiction of the human body. Conscious of the fact that the purpose of their work was to render their subjects eternal, they used firm strokes and precise contours. For example, as a human head is seen and reproduced more easily in profile, they did not hesitate in always drawing it side-on. In contrast, the eyes and organs were always painted as though seen from head-on, clearly and sharply. In other words, the eyes were always facing fully forward, while the faces were drawn in profile.

However, the upper half of the body (the shoulders and thorax) were more easily depicted from the front, as this way it was possible to draw the arms to the sides. In contrast, moving arms and feet were more clearly represented from the side. As a result, in these representations, human figures seem flat and, at the same time, twisted. Furthermore, Egyptian artists found it difficult to represent the left foot from the outside; as a result, they preferred to clearly profile it with the big toe in the foreground. Thus, both feet were shown from the side and the figure in the relief is depicted almost as if he or she had two left feet. As was the case with perspective, artists around the Nile were limited to following a rule that demanded they depict everything considered important when painting the human form. The strictness of rules also reflected a magical belief: a man drawn with his arms in profile was unable to fully undertake the task of taking or receiving the gifts that the exchange between the souls of the deceased and the gods always entailed, which formed an inherent procedure in the course of death.

Proportions

Egyptian artists started by drawing a grid of straight lines on the wall and distributing the figures over these lines with the utmost care. This grid reproduced exact measurements for each part of the body. However, this geometric sense of order did not prevent the details of different aspects of nature from being replicated with surprising accuracy. Each bird, fish, and butterfly is drawn with such faithfulness that zoologists and botanists have even been able to recognize the species depicted.

Applying the same level of strictness, seated statues had to be depicted with their hands on their knees; men had to be painted in a darker color than their female counterparts, and the predetermined representation of every deity had to be strictly respected. As a vestige of animism that precedes even the merger between Upper and Lower Egypt into a single empire, Egyptian gods were associated with an animal that artists never forgot: for example, Horus was always represented as a falcon, and Anubis as a jackal.

In a world obsessed by eternity, nobody imagined or requested a different or original form of representation. On the contrary, adhering

PORTRAIT OF NATURE
Two examples of drawings of birds in wall paintings found in the mastabas at Meidum, dated to the third and fourth Dynasties.

Artists had a completely unique way of depicting real life; they did not seek beauty, but perfection, in line with their desire to render their subjects eternal

DAILY SCENES
In frescoes and engravings, depictions of daily life could be found in abundance, such as these two men cooking and baking bread.

to this rule was an essential prerequisite for all artists. To this end, Egyptian art underwent very few changes throughout history and only slight regional variations can be seen, but the way of depicting man and nature continued exactly, in essence, the same. Only Amenhotep IV, later Akhenaten, dared to transgress these preestablished aesthetic boundaries. Allowing artists to use a more expressive or realistic style, he ordered paintings that must have stunned contemporary Egyptians. They included portraits of him kneeling beside his daughter and strolling through the garden with his wife, supported by a walking stick. Throughout his reign, some of the most beautiful sculptures of Ancient Egypt were produced, such as the famous bust of Queen Nefertiti.

SCULPTURE
Bust made from terracotta dated to the second millennium BCE, depicting a mourning woman, who could be Isis grieving over the death of Osiris.

Writing and literature

Each artist also had to learn how to write clearly, in order to engrave images and hieroglyphic symbols onto solid stone in a uniform manner. The center of written production was known as the "House of Life," and was adjacent to each temple. Here, scribes made copies of a whole range of traditional texts, including a large number of articles of a didactic nature, containing proverbs and sayings designed for young people, and religious compositions, rituals, and hymns for the gods and the Pharaohs. Although evidence exists of only a handful of examples, it is believed that narratives were also popular, in the form of stories in which reality and fiction were perfectly balanced. The best examples of such texts are *The Story of Sinuhe* and *The Story of Wenamun*, in addition to amorous and epic poems.

This tradition continued almost uninterrupted until the third century CE, although only very few texts survived to be translated from hieroglyphs

In a world obsessed by eternity, no one imagined or requested a different or original form of representation; to this end, Egyptian art underwent very few changes

to demotic or popular writing styles. Toward the end of this period, the popularity of certain literary works became more widespread. However, they were banned and destroyed by the Catholic Church when Christianity became the official religion of Rome. Egyptian demotic writing, which corresponds to the language spoken between the seventh and sixth centuries BCE, superseded Egyptian from the Late Kingdom, which was preserved as the official writing style.

The chasm between written and spoken language gradually grew. During the second century CE, certain Egyptian texts on the topic of magic were written in Greek characters. From the fourth century onward, this fusion gave rise to the Coptic language, which would become the language of Egyptian Christians, and which gradually surrendered to Arabic after 640 CE.

Arts and industries

Craft production was predominantly targeted at the upper classes, official institutions, and temples. Producing fine objects required a long period of training, which began during adolescence. Master artisans who worked for the elite were educated in temples and acquired a high social status.

Therefore, crafts occupied a prime position among the productive activities of Ancient Egypt. Evidence of this can be seen in reliefs and paintings found in tombs; scenes depicting the production of objects by a craftsman are quite common. These representations transmit the feeling of a great technical ability. For example, there are various drawings of goldsmiths weighing gold; shipowners building the hull of a boat; men melting metal; and carpenters putting the final touches to wooden symbols at a sanctuary.

HANDICRAFTS
Jars for fragrances dated to between 1550 and 1300 BCE. Egyptian handicrafts involved a great level of refinement and technical skill.

Keys of Egyptian Art

During the Old Kingdom, the rules that would govern the entire history of Egyptian art were established. Despite being in constant evolution, the canons and standards maintained were unchangeable. The periods of greatest artistic splendor coincided with the periods in which the empire reached its peak.

ANONYMOUS ARTISTS
Egyptian artists remained anonymous as craftspeople who adhered to preestablished rules. They were not expected to be original, but to implement orders with precision.

GLORIFICATION OF POWER
Art was at the service of the Pharaoh, who was immortalized in colossal sculptures, wall paintings, and reliefs.

Practicality

Egyptian art as it is known today was not conceived as such in Ancient Egypt. Its funerary monuments and temples, sculptures, reliefs, and paintings were usually the fruits of a collective work and served a practical purpose, often linked to religion or daily use.

MATERIALS
Artistic expressions were dependent on geographical location. Materials that were within reach were most often used: in architecture, large granite blocks were integrated perfectly into the surroundings.

Precursor of art

The balance sought in Egyptian art and the concept of beauty make it more similar to Eastern art than its Western counterpart. It is considered a precursor to Greek art.

Gods. Gold and lapis lazuli statuettes depicting the gods Isis, Osiris, and Horus.

ETERNITY
The full depiction of the Pharaoh at the temple followed the rules of front perspective, which helped to enhance the sense of calm and power. His expression is serene, showing no emotion, with his gaze fixed on eternity.

Ramesses II. Seated statue of Ramesses II at the Temple of Luxor.

Eternity

Egyptian art was conceived to be long lasting and eternal. The use of sturdy materials, in geometric shapes, with straight rather than curved lines, on enormous yet simple supports, made from large stone blocks, as below, were key factors in ensuring its creators successfully achieved this end.

Grandeur. The huge sizes of Egyptian constructions symbolized the power of the sovereigns.

Symbolism

Art serves to support the iconography of religion and power. The image is loaded with symbolism and has no aesthetic value. Its purpose is to transmit the message as clearly as possible. To this end, a conceptual image is used in favor of a perspective image.

Tombs. The Pharaohs were the artists' biggest customers. Their tombs were decorated with wall paintings that represented them on their journey to the afterlife.

Architecture

Ancient Egyptian architecture is the civilization's most spectacular and enduring artistic expression. The simplicity of civil buildings, made from adobe, contrasts with the grandeur and beauty of large stone constructions—the funerary monuments and temples erected in honor of the Pharaoh and the gods.

Age of royal tombs

The most important constructions from the Old Kingdom were the royal tombs. Buildings would evolve from the initial mastabas to the great pyramids and funerary temples.

Pyramid and mastaba belonging to Djoser. Third Dynasty.

Middle Kingdom constructions

During the Middle Kingdom, there was a shift from using adobe to build temples toward a preference for stone. Among the buildings preserved from the period, the White Chapel of Senusret I at Karnak is worth particular mention. The great pyramids of the Old Kingdom gave way to hypogeums for members of the royal family. The fact that they were buried underground made it more difficult for looters to spot them.

DECORATION
The walls and pillars of this small, white limestone sanctuary are decorated with bas-reliefs of the king, along with deities and hieroglyphs.

RECONSTRUCTION
The White Chapel was rebuilt in the open-air museum at Karnak. It had been dismantled and used as filler material for the construction of a pylon ordered by Amenhotep III.

White Chapel of Senusret I. Erected for the Sed festival, it is one of the most distinguished constructions from the Middle Kingdom, given its elegance and harmony.

Temple of Luxor. Erected at the request of Amenhotep III, it was later expanded by Ramesses II, who added the first courtyard, the statues and a pylon (see above), preceded by two colossal statues and two obelisks, of which only one remains.

The great temples

The New Kingdom saw the greatest splendor in this period of architecture. Great temples dedicated to the gods were erected; among these, the Temple of Luxor, dedicated to Amun, is worth particular mention, as is the Temple of Seti I at Abydos, those at Abu Simbel, and the Temple of Amun at Karnak, among many others.

GRANDEUR
The pylon erected during the reign of Ramesses II preceded a courtyard with columns, the antechamber of the imposing temple of Amenhotep III.

Structure of the temples

After the changes witnessed during the Old and Middle Kingdoms, the structure of the temples during the New Kingdom became more standardized and, starting during the reign of Amenhotep III, a floor plan was adopted that allowed for certain variations. This is the basic structure:

AVENUE OF SPHINXES AND PYLONS
The sphinxes line the avenue that runs to the main entrance of the temple, the gate to which is flanked by two obelisks. In front of the pylon, colossal statues of the king who ordered the construction of the temple rise from the ground.

PATIO
With arcades along the sides and the far end, the patio is in front of the hypostyle hall, whose floor is at a higher level.

HYPOSTYLE HALL
Features papyriform columns, with those in the center higher than those to the sides. It is usually complemented with annex rooms, such as a boat sanctuary.

CHAPEL OF GOD
The main room of the closed temple was inaccessible to anyone not linked to the cult. It contained a small statue of the god, in addition to chapels for other deities, treasures, and the like.

The Palace at Amarna

During the mid-fourteenth century BCE, Akhenaten founded the new capital, Akhetaten, in the desert. Among the temples, administrative buildings, and residences, he erected a palace to the north of the city with spacious gardens and symbols of nature. The site, today in ruins, still amazes Egyptologists.

GREEN ROOM
Decorated with a continuous frieze that reflected the nature of the region. There is also a series of niches that may have contained birdhouses for certain species.

STABLES
Mountain goats and gazelles were the predominant animals here.

POND OR POOL
They were usually surrounded by thick vegetation that may have included palm trees and other large trees.

First courtyard

SOLAR TEMPLE
Dedicated to the god Aten, center of the monotheist cult created by Akhenaten. It featured three altars, the bases of which can still be seen today.

SOUTH SECTION
The service staff and guards may have lived here.

PYLONS
This pyramid-shaped construction was a traditional part of Egyptian palatial architecture.

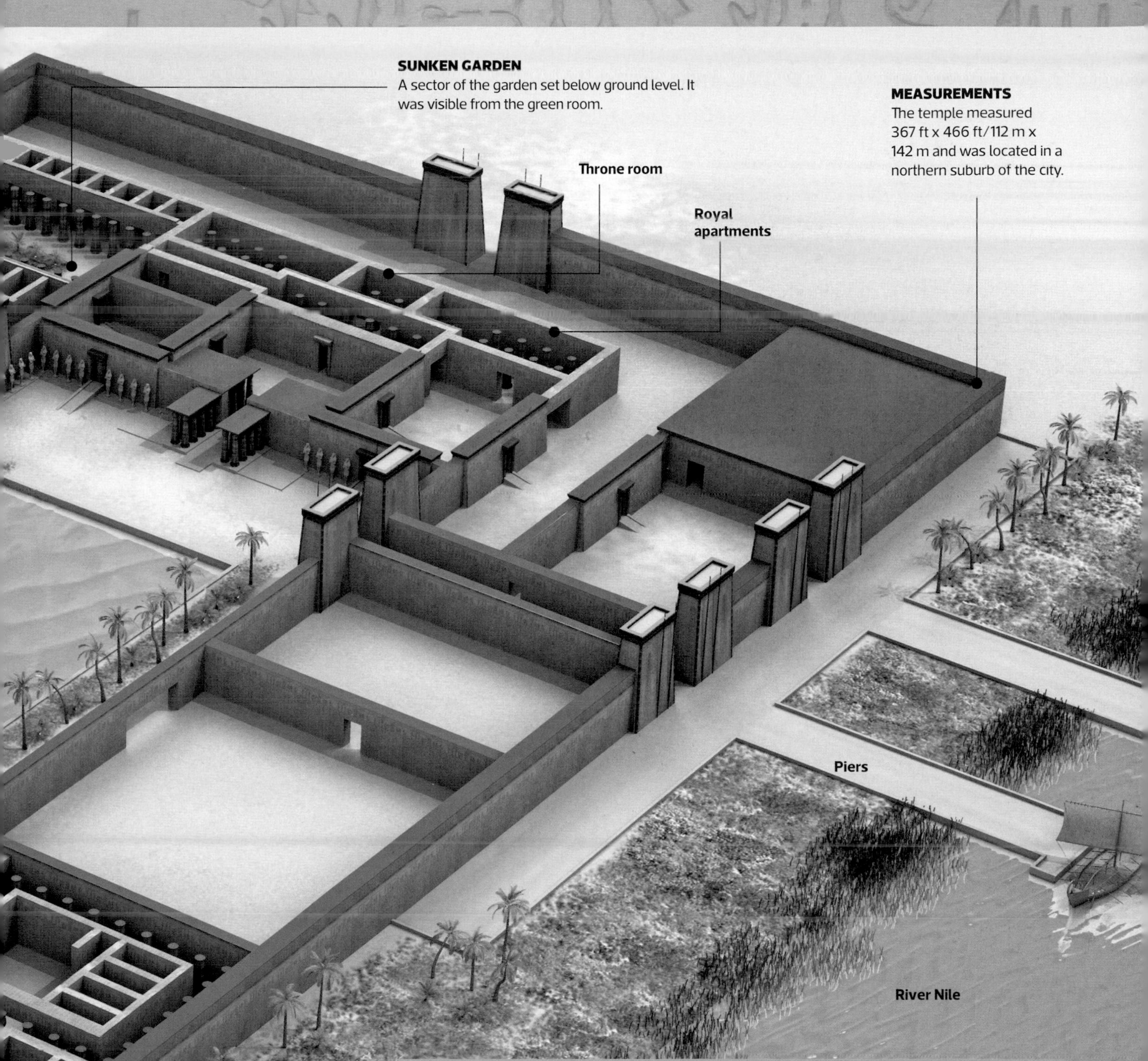

AKHETATEN (PRESENT-DAY TELL EL-AMARNA)

▶ **Period:** fourteenth century BCE. It is believed the city was not inhabited for over 20 years and was abandoned after the death of Akhenaten.

▶ **Inhabitants:** it is likely that, at its peak, the city was home to around 50,000 inhabitants.

▶ **Discovery:** during the eighteenth century.

▶ **Location:**

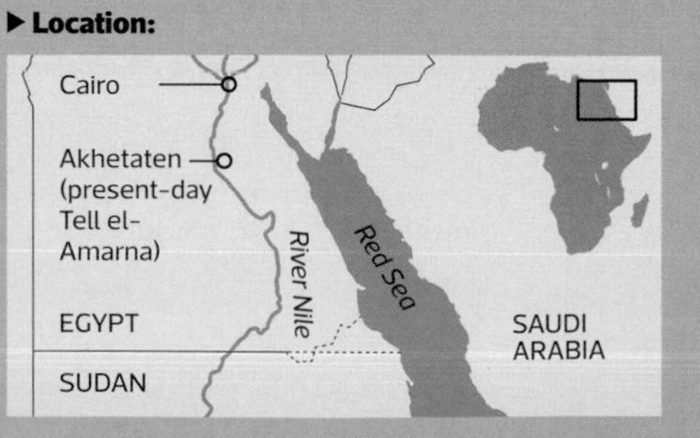

The Temple of Khonsu

Built inside the sanctuary of Amun, in the sacred complex at Karnak, the temple dedicated to the cult of the god Khonsu is a paradigm of the religious constructions of the New Kingdom and one of the best preserved.

The Theban triad

Khonsu was a lunar god associated with medicine and was the adopted son of Amun and Mut; along with his parents, he featured in the so-called Theban Triad. Construction of the temple began during the reign of Ramesses III, but it was progressively modified and expanded until the Ptolomaic Period.

The Karnak complex

Comprising three sanctuaries: one dedicated to Amun, the universal god; another to Mut, his wife; and the third to Montu, the god of war. Located to the north of Thebes, construction was started during the eleventh Dynasty in 2134 BCE, but it would not take on its definitive shape until the New Kingdom.

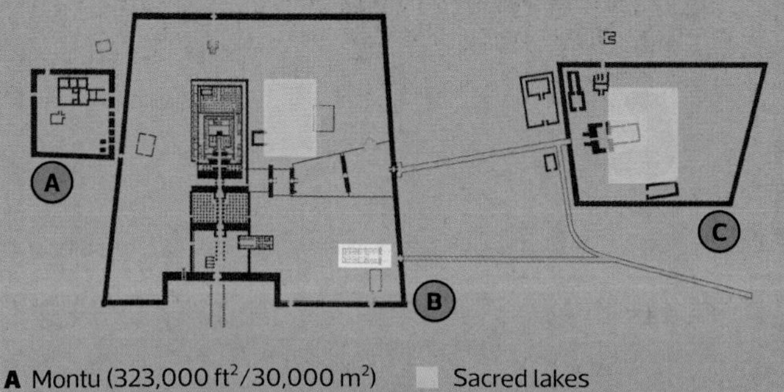

A Montu (323,000 ft²/30,000 m²)
B Amun (3,014,000 ft²/280,000 m²)
C Mut (969,000 ft²/90,000 m²)

Sacred lakes
Temple of Khonsu

ENTRANCE
An avenue of sphinxes marked the path to the temple where, in this instance, pennants indicated the location of the entrance to the House of God.

THE PYLON
The main entrance was marked out by two large trapezoidal blocks. They included internal steps to gain access to the walkway above the door.

THE COURTYARD
Open to the elements and flanked by two rows of columns. This was the public part, and even laymen could enter. It featured an altar for offerings and religious sacrifices.

Nine centuries of decline

Roman domination marked the beginning of Karnak's decline, with some of its elements being transferred to Rome stone by stone. The later use of a temple at the complex as a church did not stop its decline. After the expulsion of the Christians in the sixth century, the complex was used as a quarry until the mid-nineteenth century. It was then that interest in the ruins was reignited, with cleaning and restoration work beginning in 1905.

BOAT ROOM
Chapel used to house the sacred boat that carried the sculpture of Khonsu during the festival procession. This location was under complete darkness.

HEIGHT
Moving farther into the temple, the roofs become lower and the floors higher. This effect created the appropriate sense of seclusion for the cult.

THE SANCTUARY
Found in the deepest and darkest corner of the sacred complex. Access was exclusive to the Pharaoh or the high priest. A sculpture of the god was housed on the altar.

THE HYPOSTYLE HALL
Held up by columns that looked like a forest. Only the Pharaoh, his family, and the priests could access this covered space lit by the light from the courtyard.

The columns

Crafted from basalt, granite, calcareous rock, and alabaster. The shafts were decorated with paintings and reliefs with motifs depicting nature and religious scenes. The capitals imitated Nile vegetation or the head of the goddess Hathor.

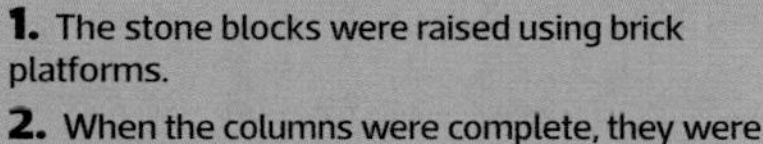

1. The stone blocks were raised using brick platforms.
2. When the columns were complete, they were shaped and the platforms were removed.
3. Finally, scaffolding was erected to decorate them with paintings and reliefs.

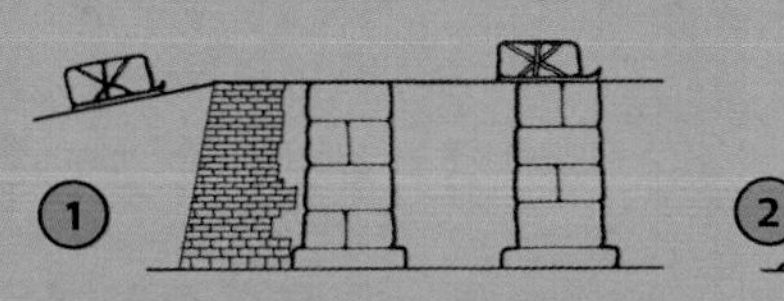

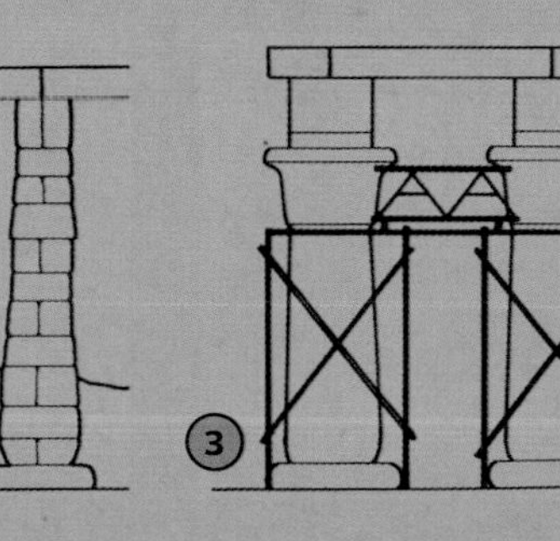

Noteworthy Pyramids

The pyramids represent the greatest expression of architecture from the Old Kingdom, in addition to being the fundamental part of the architectural ensemble dedicated to the cult of the Pharaoh. Many were not used as tombs, despite this being their theoretical purpose.

Giza plateau

Although Egypt has at least 120 pyramids, the three largest at the Giza necropolis are without doubt the most famous. They are the pyramids of Khufu, Khafre, and Menkaure.

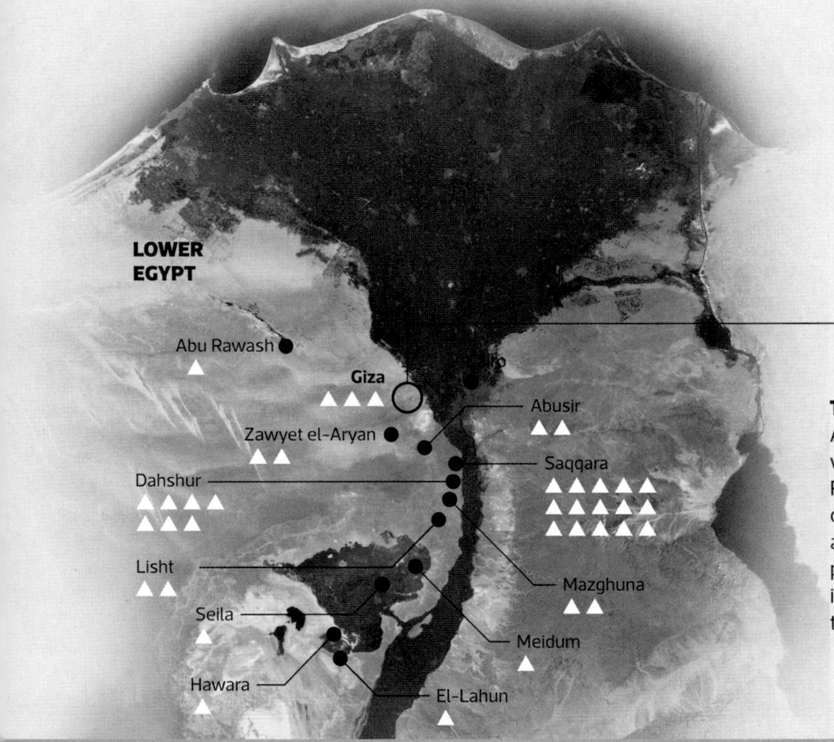

LOWER EGYPT

Abu Rawash

Giza

Cairo

Abusir

Zawyet el-Aryan

Saqqara

Dahshur

Lisht

Mazghuna

Seila

Meidum

Hawara

El-Lahun

Expanded area

Hemon's mastaba

Hypogeums

Artisans' storage rooms

Pyramid of Menkaure

Pyramids of the queens of Menkaure

THE MAIN PYRAMIDS
All can be found on the western bank of the River Nile, the place of the dead, close to a settlement which probably guaranteed the infrastructure required for their construction.

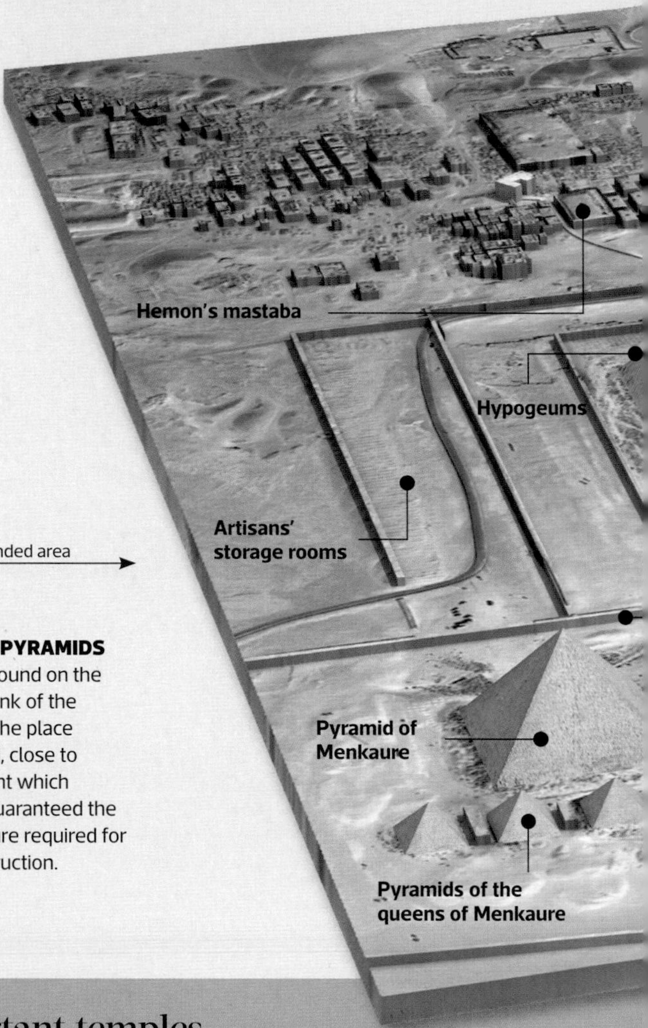

Chronology of the construction of the most important temples

PYRAMID OF DJOSER
The oldest ancient pyramid preserved to this day. It is located in Saqqara.

PYRAMID OF MEIDUM
The first erected with a false arch. Built in Faiyum, 62 miles/100 km from Cairo.

BENT PYRAMID
Its construction was ordered by Sneferu in Dahshur. It is considered an intermediate phase between "stepped" and classic pyramids.

RED PYRAMID
Also erected by Sneferu. It is believed to be the first pyramid built featuring straight sides.

PYRAMID OF KHUFU
Built by Khufu in Giza, it is the largest pyramid in the world.

PYRAMID OF KHAFRE
Located next to the Pyramid of Khufu, it was built by Khafre, who ordered the construction of the Great Sphinx.

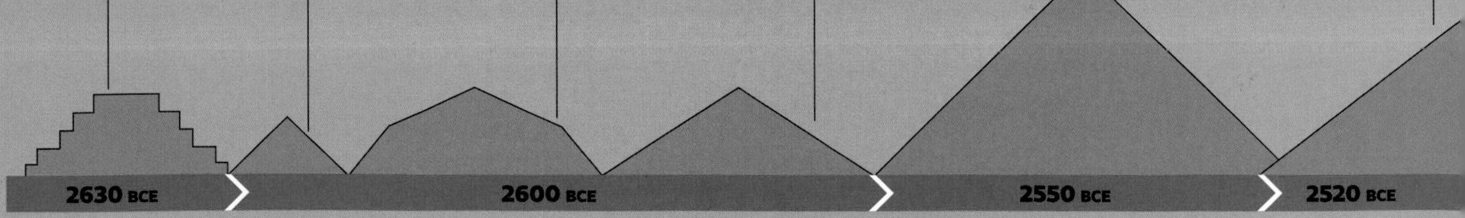

2630 BCE

2600 BCE

2550 BCE

2520 BCE

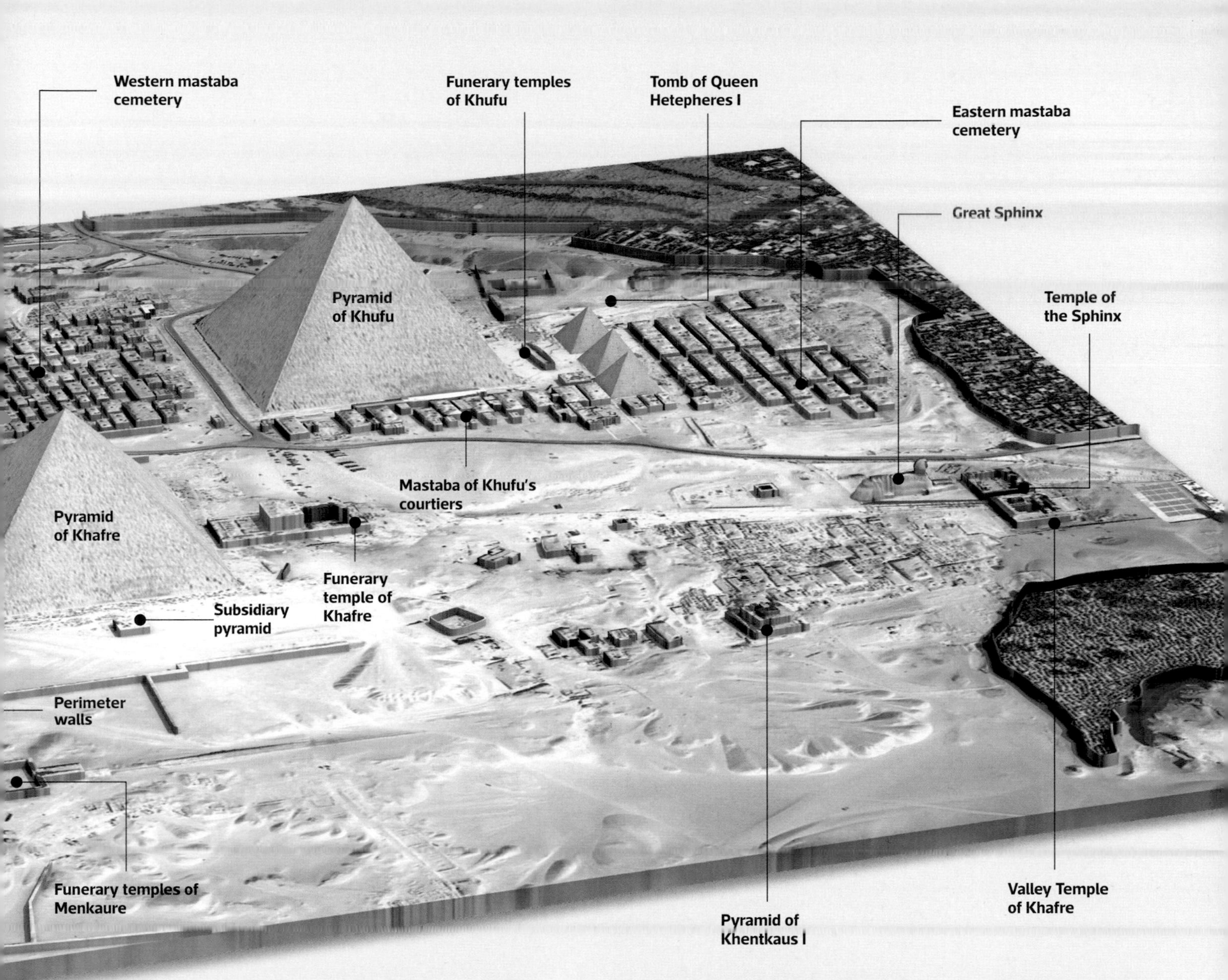

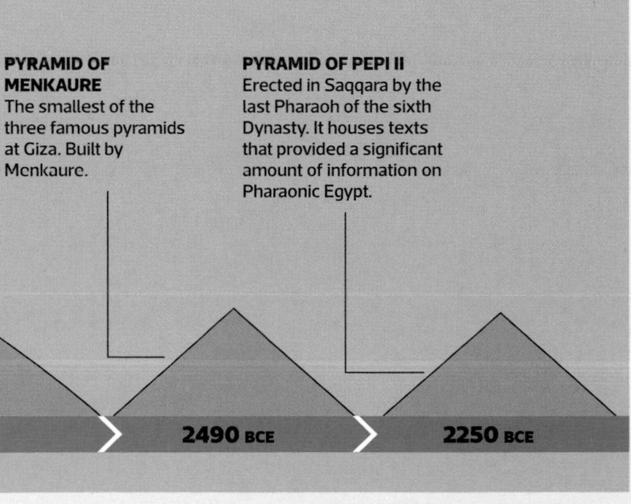

Three types of pyramid

Egyptian pyramids can be divided into three categories, depending on their shape:

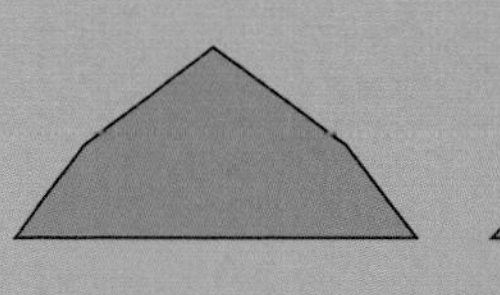

BENT
The double inclination might be attributable to a calculation error, corrected during construction to prevent it from collapsing.

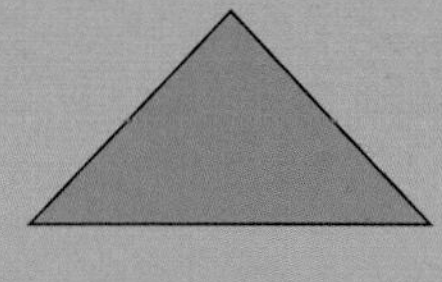

STRAIGHT
From a square base, the classic pyramid stretches uniformly up to the peak.

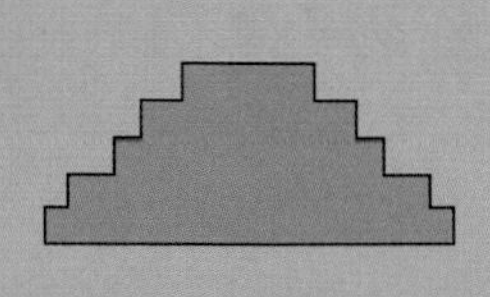

STEPPED
Comprising six superimposed rectangular-based pyramidal frustums, which decrease in size to the peak.

The First Pyramids

Primitive pyramids were erected during the third Dynasty. They were stepped to help the ascension of the king toward the sun god. The most monumental preserved to this day belongs to Djoser, constructed in 2630 BCE.

Imhotep

The architect responsible for the Pyramid of Djoser was also the royal vizier. Considered the founder of Egyptian medicine and the first known architect in history, he was adored as a deity by Egyptians and Greeks, who adopted him as Asclepius, god of wisdom and medicine.

The Pyramid of Djoser

The first pyramid ever built and one of the oldest carved stone monuments in history. It served as the prototype for pyramids that followed it. During the following years, Pharaohs like Sekhemkhet and Khaba also built stepped pyramids. With the pyramids in Meidum and Dahshur, the first king of the fourth Dynasty, Sneferu, perfected pyramid architecture.

SIZE
It measures 197 ft/ 60 m in height, and the base measures 410 ft x 358 ft/ 125 m x 109 m.

TIERS
Comprising six huge superimposed platforms, the size of which decreases the farther they are from the ground.

LOOTING
The pyramid of Djoser was ransacked by looters in ancient times, although thousands of jars were left in its galleries.

FUNERARY CHAMBER
Beneath the pyramid, the funerary chamber of Djoser was carved out of the rock.

THE MASTABAS
Prior to the construction of the Pyramid of Djoser, royal tombs in Ancient Egypt comprised underground chambers covered by an adobe superstructure in the form of a truncated pyramid. They are known as "mastabas."

THE "FALSE PYRAMID" AT MEIDUM
It was the first nonstepped Egyptian pyramid with an internal funerary chamber (rather than having been carved into the rock). Initially it was stepped; however, Sneferu later ordered the construction of straight faces. Only its internal structure has been preserved.

THE BENT PYRAMID AT DAHSHUR
Also known as the "double-shaped" pyramid, this is another pyramid erected by Sneferu characterized by its rhomboidal shape. It was meant to exceed 427 ft/130 m in height, but topped off at 345 ft/105 m due to a change in the slope of its faces during construction.

Recreation of the Djoser complex at Saqqara, the necropolis of Memphis.

Djoser. The Pharaoh who ordered the construction of the first pyramid is depicted with a braided beard and *nemes* (headdress).

The Saqqara complex

Surrounded by large, 33-ft/10-m high walls. In addition to the stepped pyramid, it housed a large courtyard, a temple designed for the funerary cult of the king, and different small-scale reproductions of the facades of symbolic buildings dating to the Old Kingdom.

CARVED STONE
The architect Imhotep introduced the use of stone, rather than brick, as the building material for this pyramid in Ancient Egypt.

The Great Pyramid

Khufu ordered the construction of the Great Pyramid around 2550 BCE. It is located at the heart of a complex containing religious buildings. Even though its entrance was hidden, it was looted in 2150 BCE.

Roof of the World

The Great Pyramid was the highest building on the planet until the construction of the Eiffel Tower in 1889. It comprises 2.3 million stone blocks, each of which weighs 2.5 tons on average.

HEIGHT
The original height of the Great Pyramid, 479 ft/146 m, has been reduced to its current 449 ft/137 m as a consequence of erosion.

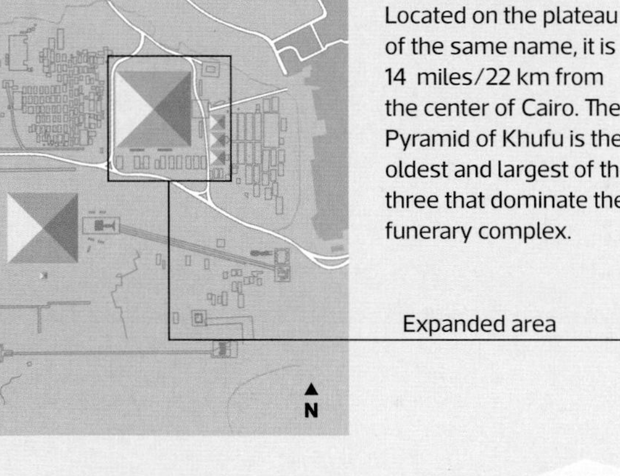

GIZA NECROPOLIS
Located on the plateau of the same name, it is 14 miles/22 km from the center of Cairo. The Pyramid of Khufu is the oldest and largest of the three that dominate the funerary complex.

Expanded area

479 ft/146 m

449 ft/ 137 m

230 ft/ 70 m

164 ft/ 50 m

0 ft/ 0 m

-98 ft/ -30 m

Ventilation ducts

The queen's chamber

Underground chamber

Mastabas

FUNERARY TEMPLE
The place where offerings were made.

Pyramids of the queens

Pits for solar boats

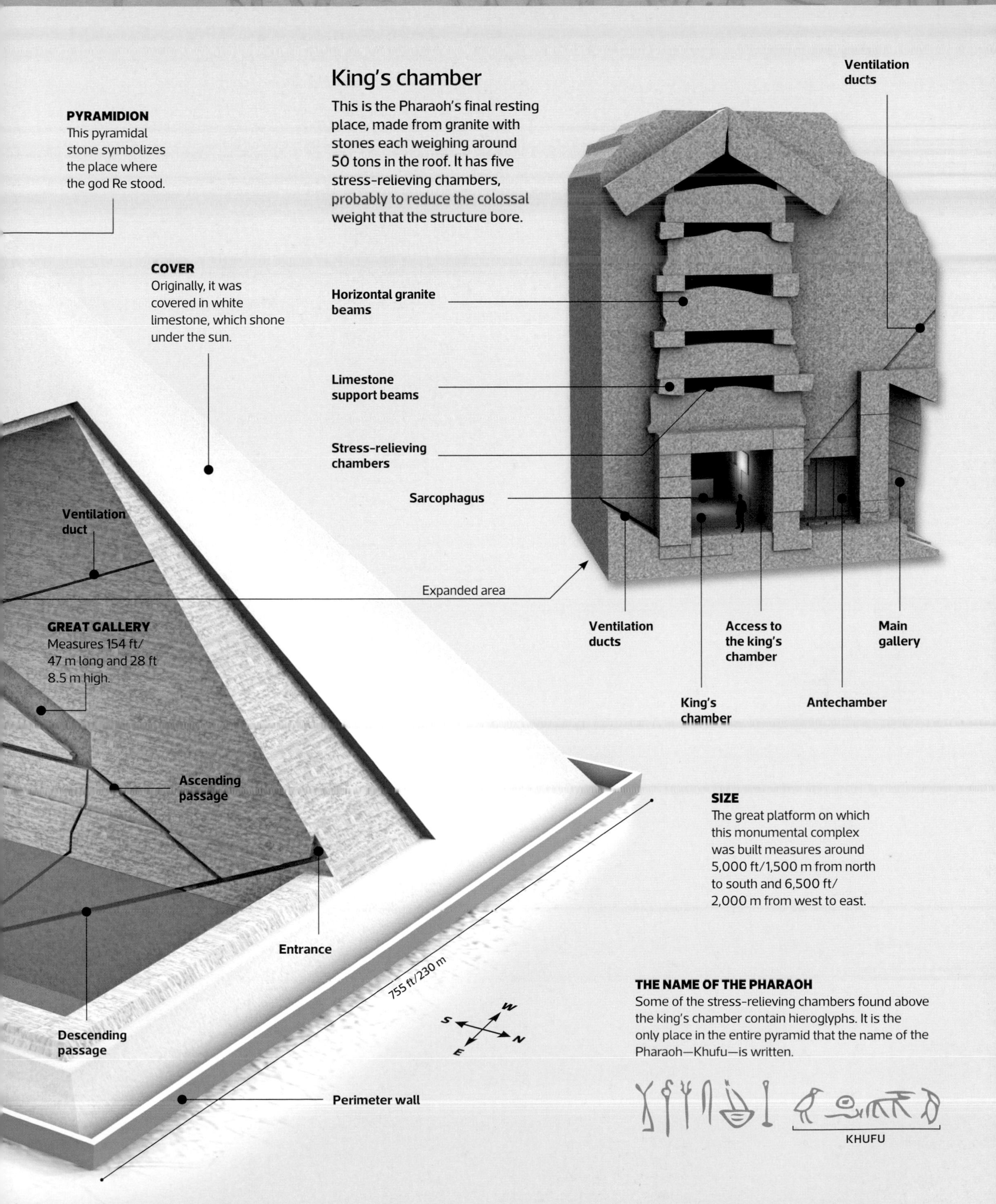

King's chamber
This is the Pharaoh's final resting place, made from granite with stones each weighing around 50 tons in the roof. It has five stress-relieving chambers, probably to reduce the colossal weight that the structure bore.
PYRAMIDION
This pyramidal stone symbolizes the place where the god Re stood.
COVER
Originally, it was covered in white limestone, which shone under the sun.
Ventilation ducts
Horizontal granite beams
Limestone support beams
Stress-relieving chambers
Sarcophagus
Ventilation duct
Expanded area
GREAT GALLERY
Measures 154 ft/ 47 m long and 28 ft 8.5 m high.
Ventilation ducts
Access to the king's chamber
Main gallery
King's chamber
Antechamber
Ascending passage
Entrance
Descending passage
Perimeter wall
755 ft/230 m
W
S
N
E
SIZE
The great platform on which this monumental complex was built measures around 5,000 ft/1,500 m from north to south and 6,500 ft/ 2,000 m from west to east.
THE NAME OF THE PHARAOH
Some of the stress-relieving chambers found above the king's chamber contain hieroglyphs. It is the only place in the entire pyramid that the name of the Pharaoh—Khufu—is written.
KHUFU

Great Builders

To erect such monumental structures as the pyramids of Giza, efficient planning and a constant supply of resources were essential to ensure they were completed while the Pharaoh was still alive.

Stone blocks

Irregular in size, and less heavy the closer they are to the summit. Two blocks of average weight (2.5 tons each) are equivalent in mass to an African elephant.

Mysterious process

Egyptologists are still looking for conclusive answers to the mysteries that persist regarding the construction of the pyramids. However, most agree that the process would have followed these steps: 1) Leveling the soil; 2) Aligning the building, using astronomical knowledge; 3) Construction of the underground chamber (carved out from the rocky subsoil); 4) Raising the ramps to transport the stones and erect the monument.

THE RAMPS
Archaeological evidence from some pyramids suggests that ramps were used to take the stone blocks up. However, it is not clear whether a single straight ramp was used, or several spiral-shaped ramps.

Expanded area

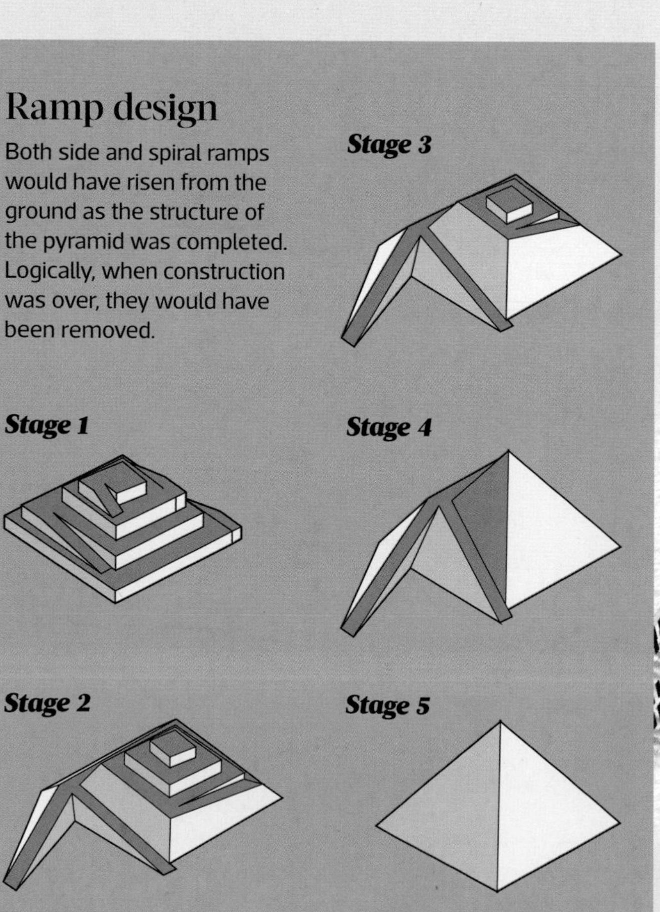

Ramp design

Both side and spiral ramps would have risen from the ground as the structure of the pyramid was completed. Logically, when construction was over, they would have been removed.

Origin of the stones

It is believed that the majority of the stones used for the pyramid of Khufu were cut in a neighboring quarry, while the rest were transported by boat along the Nile from other locations.

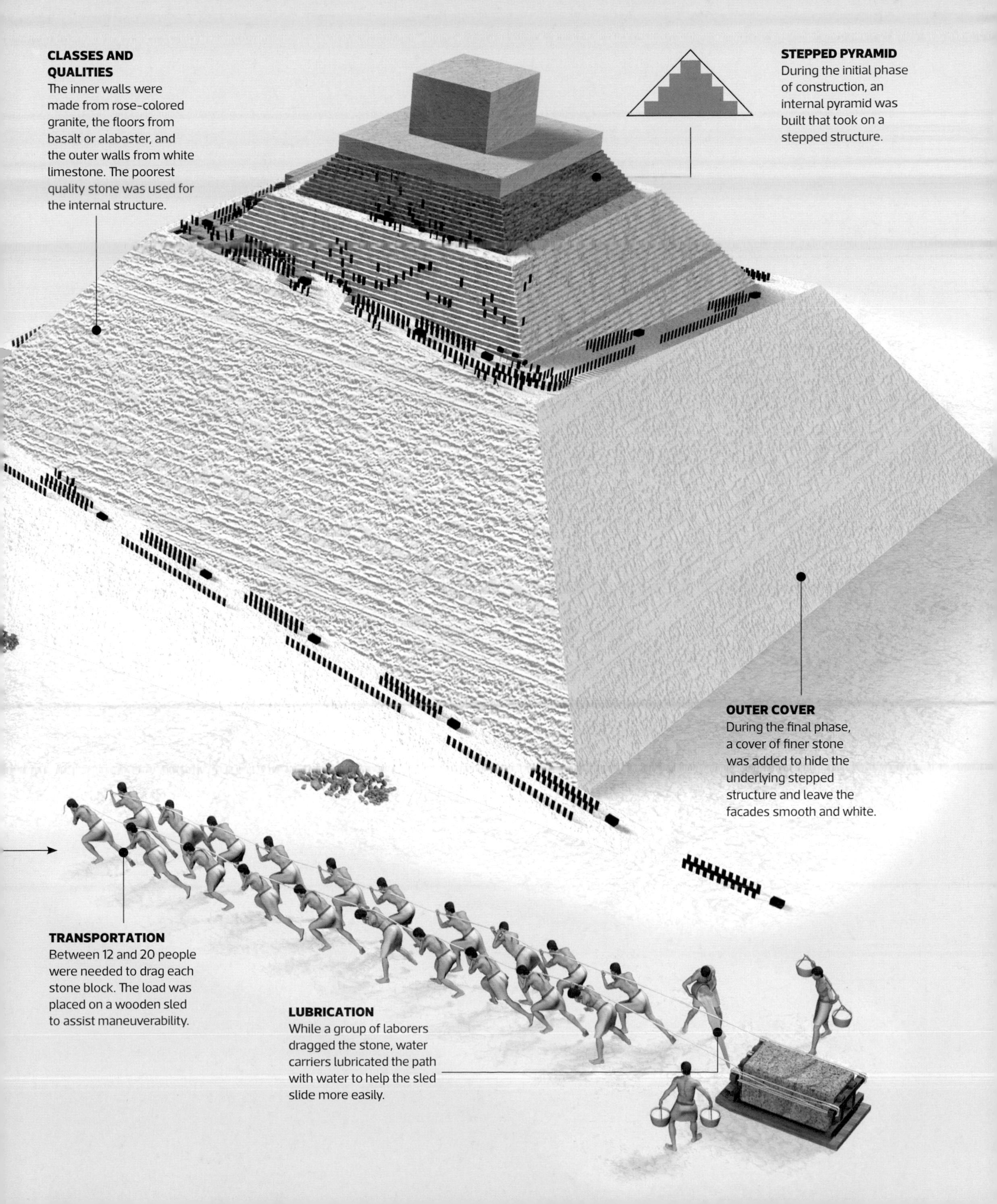

CLASSES AND QUALITIES
The inner walls were made from rose-colored granite, the floors from basalt or alabaster, and the outer walls from white limestone. The poorest quality stone was used for the internal structure.

STEPPED PYRAMID
During the initial phase of construction, an internal pyramid was built that took on a stepped structure.

OUTER COVER
During the final phase, a cover of finer stone was added to hide the underlying stepped structure and leave the facades smooth and white.

TRANSPORTATION
Between 12 and 20 people were needed to drag each stone block. The load was placed on a wooden sled to assist maneuverability.

LUBRICATION
While a group of laborers dragged the stone, water carriers lubricated the path with water to help the sled slide more easily.

The Great Sphinx

The Great Sphinx rises from the sand to the east of the pyramid of Khufu; it is the oldest and greatest figure sculpted by humankind preserved to this day. It is believed that this colossal figure of a lion with the head of a human represents Khafre.

NOSELESS
Arabian chronicles from the fifteenth century explain that the face of the Great Sphinx was damaged by the Mamluks of the period.

SIZES

▶**Construction date:**
between 2600–2500 BCE.

46 ft/14 m

66 ft/20 m

240 ft/73 m

The constructor

The stele of Thutmose IV, or the Dream Stele, is a 7-ft/2.15-m high granite slab that was erected between the figure's front paws. It mentions Khafre as having ordered the construction of the Sphinx, although this part of the text is incomplete (the stone is partially damaged).

The Sphinx's beard

As part of their restoration of the ancient glories of Egypt, the kings of the eighteenth Dynasty tackled the Great Sphinx. They installed protective layers of masonry and adorned it with the braided beard typical of the deities. However, the beard quickly fell off.

It is preserved in the British Museum in London.

LOCATION
The Sphinx is positioned in a deep pit, to the far east of the elevated walkway that leads to the pyramid of Khafre.

COLOR
It is believed that during the Old Kingdom, the body and face were painted red.

HEADDRESS
The *nemes*, or Egyptian headdress, was initially painted in horizontal blue and yellow colored stripes.

Guard and protector

Sculpted from limestone from the quarry of the Great Pyramid. The head, supposedly a portrait of Khafre, was carved from a harder stratum of rock, while the body was sculpted from other softer rocks and as a result has suffered more erosion. For centuries, it lay covered by the desert sand. Conceived as a guardian, it would have served as a guard and protector of the Giza complex.

The Sphinx, as seen by the Western world

1615	1681	1755	1798	1838	1858	1887	1925
George Sandys	**Cornelis de Bruijn**	**F. Louis Norden**	**Vivant Denon**	**David Roberts**	**Francis Frith**	**Henri Béchard**	**Émile Baraize**

The Temple at Abu Simbel

Famed for the four enormous statues of Ramesses II that feature on its facade, the temple was built during the thirteenth century BCE to commemorate the Egyptian victory at the Battle of Kadesh and in honor of the Pharaoh himself. Carved out of the rock, it was located next to the Nile, in Nubia in southern Egypt.

RELOCATION
In 1968, the temple was moved to a nearby site next to Lake Nasser, to prevent it from being submerged following the construction of the Aswan Dam.

PYLON
The four colossal statues stand out as part of the facade. They measure 115 ft/35 m wide and 98 ft/30 m high.

GIANTS
Sculpted directly into the rock, they represent Ramesses II. They measure 72 ft/22 m in height.

Size compared to the height of a human.

SMALLER FIGURES
They are located at the sides of the giants and at their feet. The first represent the Pharaoh's relatives and the second, prisoners of war.

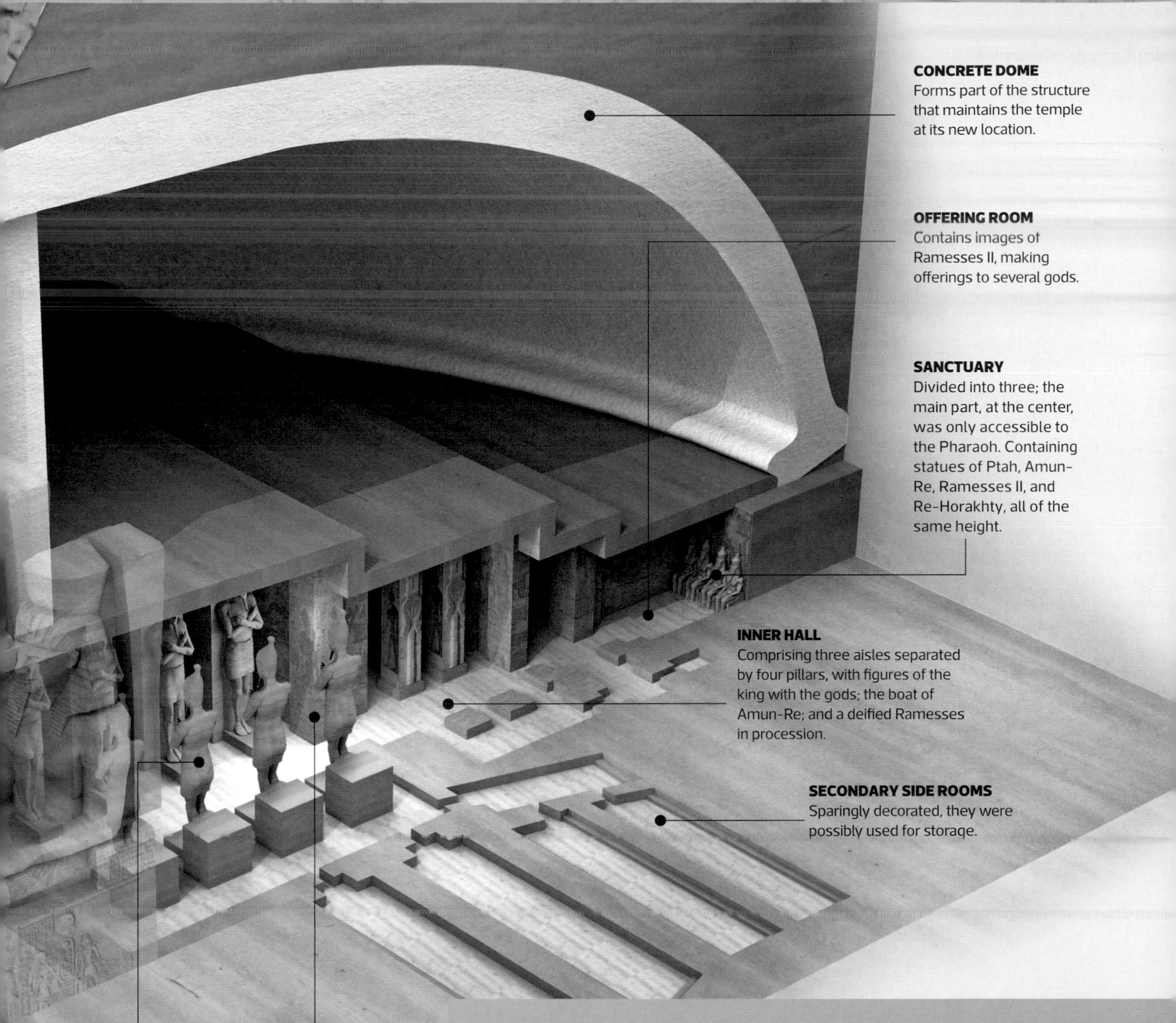

CONCRETE DOME
Forms part of the structure that maintains the temple at its new location.

OFFERING ROOM
Contains images of Ramesses II, making offerings to several gods.

SANCTUARY
Divided into three; the main part, at the center, was only accessible to the Pharaoh. Containing statues of Ptah, Amun-Re, Ramesses II, and Re-Horakhty, all of the same height.

INNER HALL
Comprising three aisles separated by four pillars, with figures of the king with the gods; the boat of Amun-Re; and a deified Ramesses in procession.

SECONDARY SIDE ROOMS
Sparingly decorated, they were possibly used for storage.

WALLS
Decorated with scenes of war, with a religious undertone, they show the Egyptians defeating their enemies.

HYPOSTYLE HALL
Measures 59 x 52 ft/ 18 x 16 m. The eight pillars featuring images of Osiris, with the physical features of the Pharaoh, are particularly noteworthy; they measure 33 ft/10 m in height.

Play of light

Twice a year, the rays of the sun shone through the door. The light was projected into the main hypostyle hall, the inner hall, and the sanctuary, illuminating the four statues in the niche to the rear.

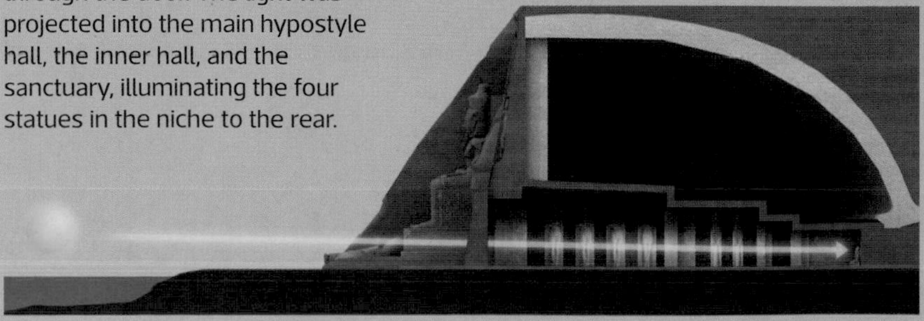

Reliefs

The reliefs that cover the facades, inner walls, and columns of temples and palaces are typical expressions of Egyptian art. Similar to paintings, scenes depicting daily life and spiritual and religious characters were frequent. Of all the techniques, bas-relief was the most common.

Evolution

The bas-relief technique was perfected during the Predynastic period. Numerous painted reliefs have been found in tombs dating back to the Old Kingdom, which stand out given their quality, variety, and dynamism. The technique spread during the Middle Kingdom, equally for civil as well as religious purposes. During the New Kingdom, the bas-reliefs at the Theban tombs are particularly noteworthy, as are the technical and aesthetic changes brought about by Akhenaten's reformations.

Akhenaten and Nefertiti. Sunk relief.

Sunk relief

This technique involved cutting models out of the material, and is typical of the Amarna period. It became more common throughout the New Kingdom. The technique made figures on open-air walls stand out, thanks to the illumination provided by the sun's rays, creating plays of light and shadows.

Bas-relief technique

As part of this technique, which was much more commonly used than others, the outlines were drawn onto the wall and the surrounding area carved out. As a result, the figures appear slightly more prominent than the background.

Polychrome bas-relief. Scene of farm workers from a private tomb from the sixth Dynasty (Saqqara).

Mid-relief

In certain reliefs, the figures are more prominent than in bas-reliefs. The background is carved out more deeply, offering a more noticeable standout effect.

FIGURES
The norms of front perspective and the proportion of figures were the same as those that ruled the art of wall paintings.

Bek and his wife, Taheret. By Bek, one of Akhenaten's sculptors.

High relief

In high reliefs, the figures stand out much more. However, this technique was quite uncommon in Ancient Egypt. The figures project by at least half their depth from the background and are sometimes considered adjoining statues.

FOUNDING RITUAL
The reliefs at the temple of Kom Ombo provide information about the different rites that the king performed in the presence of the deities.

Temple of Horus and Sobek. Ruins at Kom Ombo.

Annihilation of enemies. Eighteenth Dynasty, ca. 1450 BCE.

Royal power

This great relief depicts one of the classical scenes of Egyptian symbolism. In order to emphasize the power of the Pharaoh, the scene depicts the triumph of Thutmose III over peoples from the North. It can be found on the western face of the seventh pylon erected at Karnak, which measures more than 207 ft/63 m wide.

CRUSHING THE ENEMY
The Pharaoh, wearing the red crown of Lower Egypt and a pleated skirt, grips his enemies by their hair; they beg for mercy with their arms raised in the air.

Thutmose, the Master Sculptor

Perhaps it was the unfinished nature of the portraits by Thutmose during the reign of Akhenaten that saved them from the systematic destruction to which Amarna art was subjected. Abandoned at the sculptor's workshop, these magnificent pieces were discovered intact by Ludwig Borchardt in 1912.

Amarna sculpture

The religious reforms of Akhenaten affected all aspects of Egyptian life, including art. With an original expressive and realist style, Amarna art broke with the previous tradition. Sculptures from the end of this period are distinguished by their more delicate and harmonious features, their spontaneity and realism.

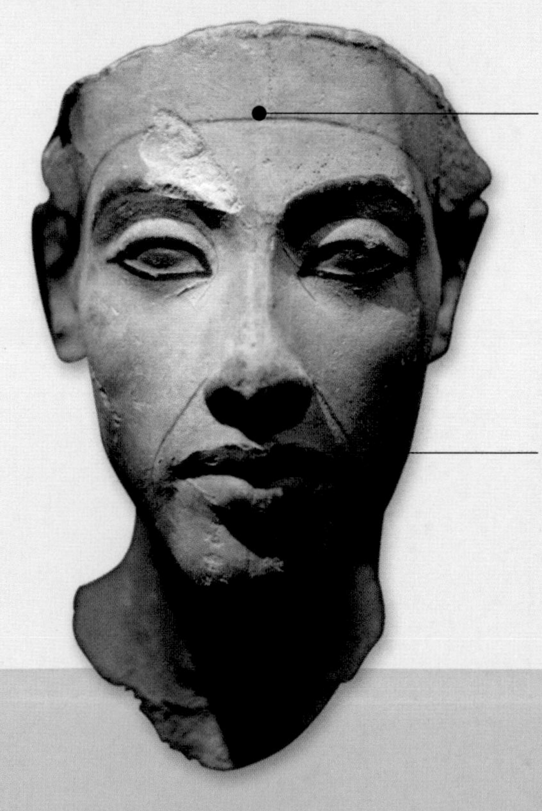

TWO HALVES
This figure is formed from two halves put together, the cast for which was taken from a finished statue and modeled using plaster.

FACIAL FEATURES
The features of Akhenaten in this portrait are much more realistic than the more exaggerated features of the first sculptures from the Amarna period.

The Pharaoh Akhenaten. Stucco sculpture measuring 10 in/25 cm in height.

Bust of Nefertiti. One of the most recognized works of Ancient Egypt, a symbol of beauty and perfection.

NEFERTITI'S HEADDRESS
Nefertiti wears a high, blue-colored crown with a diadem and a *uraeus*, with which she was usually depicted. She is also wearing a wide necklace adorned with colorful pieces.

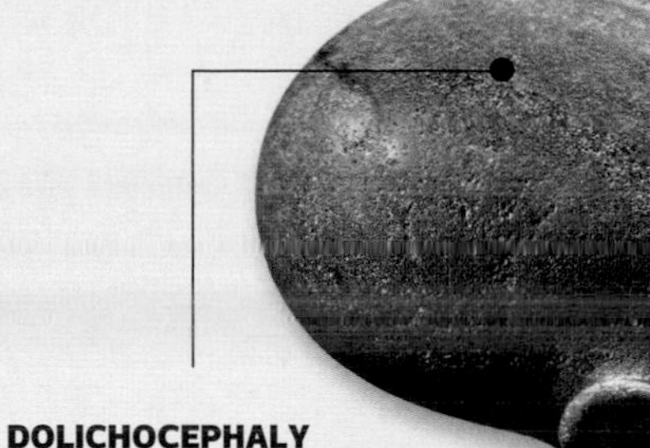

DOLICHOCEPHALY
The deformation of the head has been the source of different theories, such as a hereditary disease. However, it could also be attributed to an exaggerated representation of a feature of Akhenaten's family.

The princess

The change in Amarna art was not just stylistic; it introduced new subjects, such as the private life of the Pharaoh's family. This sculpture is one of a series of three heads of the princesses, Akhenaten's daughters, found in Thutmose's workshop. They stand out on account of their facial features, with large eyes and full lips, and extremely elongated heads.

Princess's head. Made from yellowy-brown quartzite.

UNFINISHED
The bust appears unfinished. The absence of the left eye supports the theory that this is a model that may have been used to teach apprentices at the workshop.

EPITOME OF BEAUTY
This bust is famous on account of its balance and elegance. The harmonious features conform to an ideal of timeless beauty.

Depicting old age

Depictions of old age are uncommon in Egyptian culture. These private sculptures depict an older man and an older woman. The sculptor has attempted to capture old age in the portraits in a realistic way, through the flaccidity of the skin, the lines on the foreheads, and the bags under the eyes.

MAKER
These models, and many more, were found in Thutmose's workshop and are attributed to him.

Plaster model of a man.

Plaster model of a woman.

Painting

It is impossible to break the link between paintings and reliefs, given that the decrees ruling the practices were very similar and often the reliefs were colored. Painting also appears in handcrafts and sculpture, although it was most commonly used on the walls of large temples and funerary monuments.

Preparing the wall

The most common technique for wall paintings was fresco secco, which involved preparing the wall in advance by applying plaster and allowing it to dry.

DAILY LIFE
The scenes depicted in various tombs have made it possible to understand the daily life of the Egyptian upper classes, farmers, shepherds, and artisans.

HIERARCHY
The size of the figures was directly related to their social position. Pharaohs and gods were larger in size, and other characters were depicted on a smaller scale.

Flat representations

One of the characteristics preserved throughout the history of Egyptian art is the lack of perspective. It does not feature shadows and the colors are flat. Equally in paintings and reliefs, the image has just one dimension, and tries to incorporate the maximum number of characteristics of the object represented. Many resources were used to achieve this, such as a special representation of the human body and image hierarchies.

THE GEESE OF MEIDUM
Painted with mineral colors on a plastered wall—quite an uncommon technique—it is one of the few noteworthy paintings from the Old Kingdom. Found in the tomb of Nefermaat, it is distinguished by its balanced composition and color.

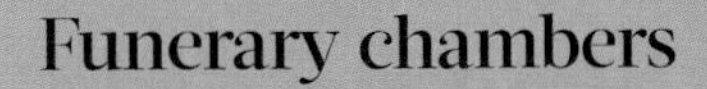

Rules on proportion

A strict system of proportions governed the representation of figures in paintings and reliefs. This system, which regulated the exact distance between body parts, was based on the use of grids or squares. The body occupied 18 squares when on foot, and 14 when seated.

RELIGIOUS TOPICS
Tomb paintings helped the Pharaoh on his journey to the afterlife. Below, the Pharaoh is accompanied by Horus and Anubis after passing the test to enter the world of the dead.

HUMAN FIGURE
Represented as a combination of several parts: the head and legs were represented in profile, with the torso and eyes seen head-on.

Funerary chambers

The walls of the tombs, where the deceased resided, feature scenes typical of the afterlife. The king, deities, and the deceased (if that were not the Pharaoh himself) were the main protagonists in these depictions.

Inside Nefertari's tomb. The viewing of the tomb paintings was reserved for the deceased.

FREEDOM
When depicting animals or other aspects of nature, a great amount of expressive freedom was allowed.

Egyptian Literature

A fundamental feature of Ancient Egyptian literary texts is that they were created to be recited or represented, rather than read as a stand-alone text. The Middle Kingdom is considered the classical period of Ancient Egyptian literature.

MUSIC
Religious hymns, love poems, prayers, and work songs were typically set to music.

Narrative

Derived from autobiographies that became common during the Old Kingdom, the masterpiece is *The Tale of Sinuhe*, dating to the Middle Kingdom. This period also saw the documentation of old folk tales, from which significant works were created. *The Tale of the Shipwrecked Sailor* and *The Story of King Cheops and the Magicians* are worth particular mention.

Relief. Scene depicting a recital accompanied by music.

The Tale of Sinuhe

Ancient Egyptians considered *The Tale of Sinuhe* their main literary classic. This narrative, in the style of an autobiography, featured high dignitary Sinuhe as its main character. In the service of Senusret I, Sinuhe discovered a plot to kill the Pharaoh's father, Amenemhat I, and fled to the east. Finally, he returned to the court to document his memoirs. The report is written in prose with parts in verse, and has significant literary and historic value.

Senusret I. Osiris pillar belonging to the Pharaoh from the twelfth Dynasty.

The "Berlin Papyri." A fragment from papyrus 3022, preserved in Berlin, discovered by Egyptologist François J. Chabas in 1863.

Religious themes

Egyptian hymns, poems dedicated to the gods, are of significant interest. This is particularly the case of the Great Hymn to Aten. Compilations of funerary texts are also worth particular mention, such as the Coffin Texts and the Book of the Dead. Although not literary compositions, they are of great literary value.

Hymn to Amun-Re. Detail of a stele from the eighteenth Dynasty, containing a hymn dedicated to the god Amun.

RECITALS
Almost always in verse. It is believed that rhymes were not used, and the rhythmic cadence was established using alliteration and other figures. Verses took the form of short couplets.

RECOVERY
The hefty text of *The Tale of Sinuhe* has been reconstructed using the so-called "Berlin Papyri." Papyri 3022 and 10499 contain most of the text, although fragments have also been found in other documents.

Mika Waltari. Prolific Finnish writer, photographed in 1939.

DISSEMINATION
The work was rescued by Finnish author Mika Waltari (1908–79) in his novel *Sinuhe, the Egyptian* (1945). Although his adaptation had very little to do with the original, it captivated readers and brought the classic work into the limelight.

Epic, politics, and teachings

Other existing texts served a more didactic purpose, containing guidance: political texts were popular during times of instability, whereas epic narrations recounted historic events such as the Battle of Kadesh and the expulsion of the Hyksos.

Love poems

During the New Kingdom, the love poem genre was created; they were always recited or sung to music. Eight collections of amorous lyrics have been preserved, containing dialogues and reflections on the difficulties of love.

Two Visions of Egypt

Ancient Egypt amazed Herodotus. The descriptions made by the Greek historian in the fifth century in his monumental *Histories* opened the door to a greater understanding of this civilization. The vision left by the Bible, including reports on Hebrew slaves in Egypt, is somewhat less flattering.

PESTILENCE
As narrated in Exodus, God warned the Pharaoh that if he did not free the people of Israel from his control, ten plagues would beset his lands. Pestilence was the fifth plague in Egypt.

Biblical Pharaohs

In the Bible, five Pharaohs were mentioned: Shishak (Shoshenq I), So (although a Pharaoh of this name was never recorded, it could refer to Tefnakht who ruled from Sais, also written as So), Taharqa, Necho II, and Hophra (Apries). Occasionally, references are made to Pharaohs without specifying their names; for example, Exodus makes reference to "the Pharaoh of the ten plagues." Hebrew slavery in Egypt was the cause of the negative opinion of the civilization.

Necho. Statuette from the seventh century BCE showing the Pharaoh making an offering to a deity.

THE OLD TESTAMENT
Exodus describes the exit of the Hebrew population from Egypt, having been held captive by the Pharaoh (whose name is not mentioned), and describes slavery, the pyramids, the court of the Pharaoh, and the "divine retribution" of the plagues.

First data

For centuries, Herodotus was the primary source of knowledge on mummification; as a secret ritual, no records of it were found in Egyptian documents. The first known description of methods used to construct the Great Pyramid also comes from Herodotus, and was considered feasible until the nineteenth century.

Herodotus' travels

Around 445 BCE, Herodotus traveled in Egypt for four months, from the mouth of the Nile to Aswan. In his *Histories* he wrote that "Egypt is the gift of the Nile," leaving records of his admiration of Egyptian art and social organization. He collected information from Greeks living there and from Egyptian priests, although it has not been ruled out that he may have based some of his descriptions on stories and legends from the time.

First historian. Born in Halicarnassus in 484 BCE, the founding father of history died in Athens in 425 BCE.

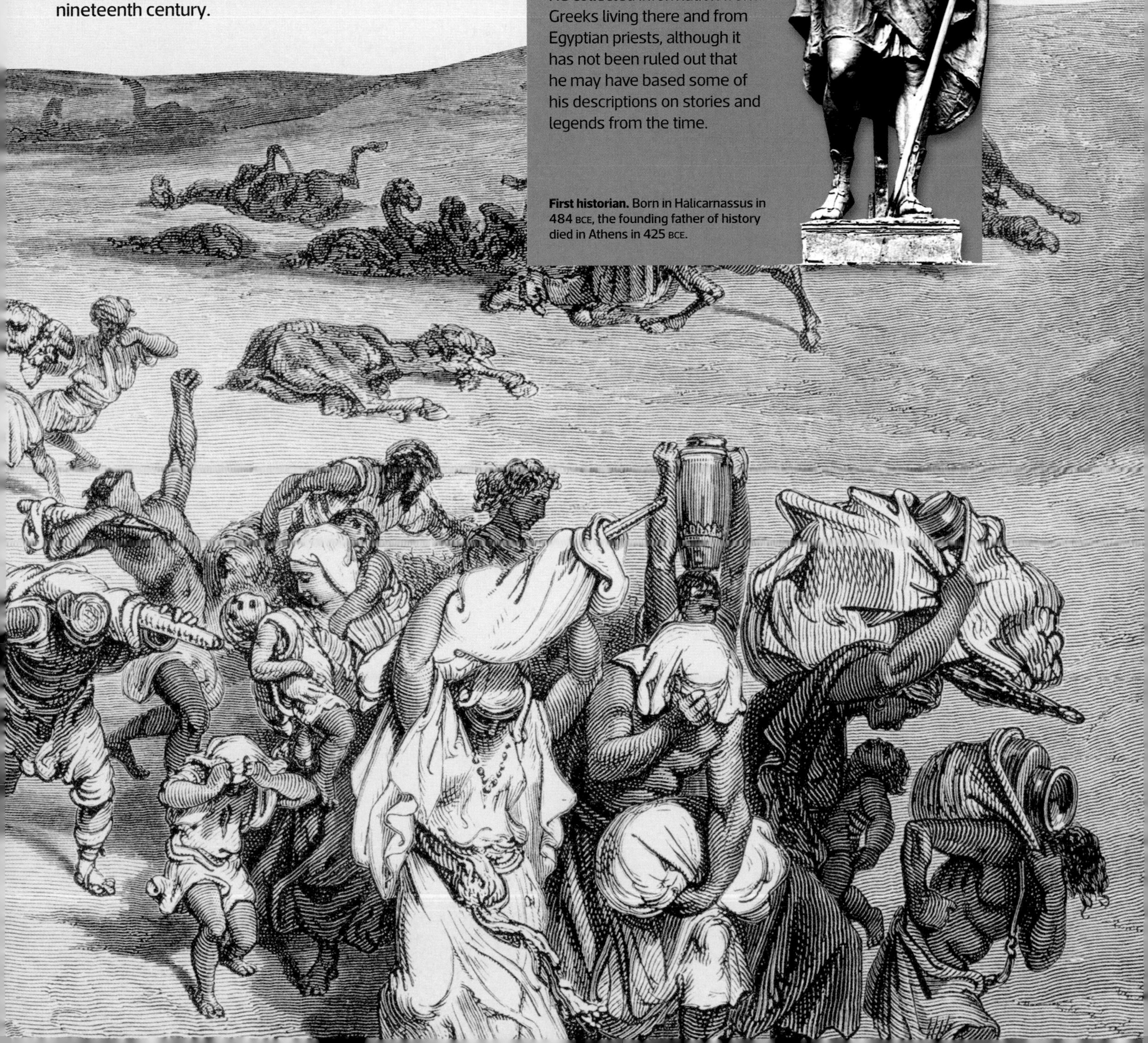

Egypt in Museums

The majority of Ancient Egyptian treasures are preserved in some of the largest museums in the world. This dispersion is attributable to the significant period during which artistic heritage was not protected, which resulted in it falling victim to archaeological looting and questionable donations.

Egyptian Museum in Turin

The main Egyptian museum in Italy preserves extremely valuable pieces from Ancient Egypt, especially from the period during which Rome controlled the region. Founded in 1824, the museum is home to 32,500 objects accumulated between the Paleolithic period and the Coptic period. Among its most precious treasures is the oldest copy ever found of the Book of the Dead, in addition to a statue of Ramesses II and the intact tomb of the illustrious architect Kha and his wife, Merit.

Ancient Egyptian amphora, dating to between 1567 and 1320 BCE.

Scattered heritage

It was not until 1983 that the Egyptian government banned the removal of objects from the country. However, it is difficult to undo the damage that Egypt has suffered since at least the era of Roman occupation. Nonetheless, the Egyptian Museum in Cairo is unquestionably the institution that is home to the largest number of objects.

The Seated Scribe. Perfectly preserved, this sculpture is considered representative of art during the Old Kingdom.

Louvre Museum in Paris

The bond that connects the Louvre to Ancient Egyptian culture was formed during the Napoleonic Campaign to the land of the Nile. A collection of more than 60,000 pieces, 5,000 of which are on display, makes it one of the museums with the most articles from this civilization. The most famous piece on display is the Seated Scribe.

Zoomorphic funerary statue of a cat (fourth century BCE), preserved at the British Museum.

British Museum in London

Cairo aside, the capital of the United Kingdom is home to the most important collection on Ancient Egypt. Among the many valuable pieces, the following are particularly noteworthy: the Rosetta Stone, which facilitated the decoding of hieroglyphs; three colossal statues of Amenhotep III and one of Ramesses II (known as "the young Memnon"); and 95 tablets known as the "Amarna letters."

The Egyptian Museum in Cairo

Its walls house the greatest collection of pieces from the Pharaonic period of Ancient Egypt, with more than 160,000 classified objects. Since 1922, the museum has experienced exponential growth, following the inclusion of Tutankhamun's funerary treasures, with more than 3,500 pieces found by Howard Carter. It is also home to the sarcophagus of Akhenaten and the statues of Ka-Aper and the seated statue of Khafre.

Mummy with head of a falcon. Dating to the first century BCE, it is one of the objects on display in the Egyptian Museum in Cairo.

Obelisks

These pillars with a pyramid-shaped top form part of the legacy that Ancient Egypt left to the world. They are often associated with the commemoration of a historic event. Public squares in various cities of the world, such as Rome, London, New York, and Paris, feature original obelisks.

Piazza del Popolo. Obelisk of Ramesses II, taken to Rome during the period of Roman control over Egypt.

Egypt on the Silver Screen

Inevitably, cinema did not escape the powerful magnetism wielded by Ancient Egypt. The enigmatic world of the Pharaohs and the scenographic potential offered by the artistic legacy of Egypt have provided ideal storylines and backdrops for the silver screen.

Recurring themes

Queen Cleopatra and tales of mummies have been two regularly recurring themes in the abundant filmography about Egypt. Since *Cléopâtre* (1889) by French filmmaker Georges Méliès, the myth of the Egyptian sovereign—with her beauty, character, and mysterious death—has seduced both directors and the public.

"CLEOPATRA" (1963)
Of all the films focusing on the figure of Cleopatra VII, without doubt the most famous was made by director Joseph L Mankiewicz, starring Elizabeth Taylor and Richard Burton. Given its high production cost, it almost bankrupted 20th Century Fox. The film studio spent many years trying to recover the investment made, despite its success at the box office.

Mummies

The spells involved in the mummification ritual have given rise to stories of mummies that, revived by incantations, came back to life. The world of film knew how to capitalize on the theme, with great hits such as *The Mummy* (1932), made by Karl Freund, and its 1999 version by Stephen Sommers.

Poster for *The Mummy*, a classic film by Karl Freund starring Boris Karloff.

Historically inaccurate

Most films on Egypt have not concentrated on historical accuracy, but instead on the spectacular nature of scenes, monumental sets, and great numbers of extras. However, there have been exceptions, such as the modest and praised *Pharaoh* (1965) by Jerzy Kawalerowicz, and *The Night of Counting the Years* (1969) by Shadi Abdel Salam.

Still from *Pharaoh*, by Academy Award winning Polish director Jerzy Kawalerowicz.

"THE TEN COMMANDMENTS" (1923 AND 1956)
Both versions were by Cecil B. DeMille. The first, a silent movie in black and white, was so successful that a remake was filmed later in color, with a greater level of historical accuracy than its predecessor. In 1934, DeMille also made *Cleopatra*.

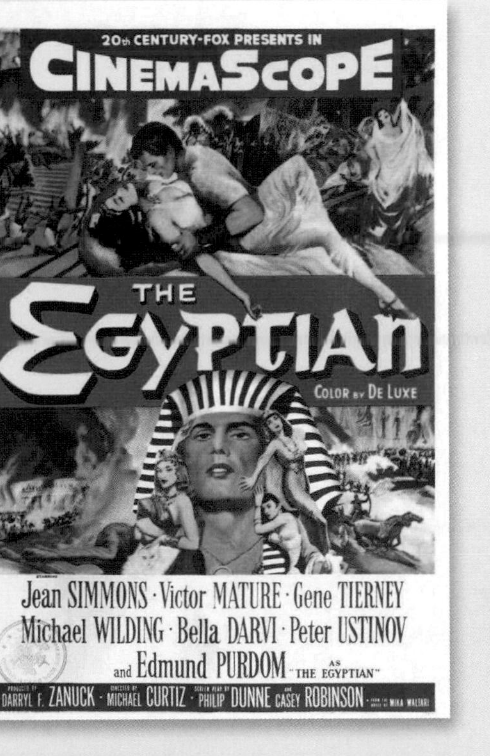

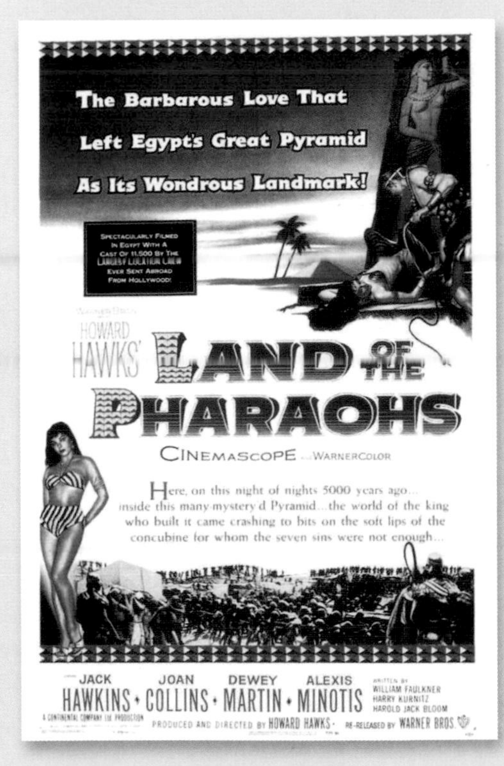

Elizabeth Taylor as Cleopatra.

"THE EGYPTIAN" (1954)
Based on the successful novel by Mika Waltari, this film made by Michael Curtiz, starring Jean Simmons and Victor Mature, is the story of Sinuhe, a physician in the court of Akhenaten. The director was unable to replicate his previous success, *Casablanca*.

"LAND OF THE PHARAOHS" (1955)
Another great director, Howard Hawks, also jumped on the Egyptian bandwagon during the decade of Cinemascope, when large-scale historical productions were at the height of popularity. This film, based on the construction of the Pyramid of Khufu, featured a strong screenplay and attractive setting.

A

B

C

D

E

F

G

I

L

M

Based on an idea by Joan Ricart

Editorial coordination Marta de la Serna

Design Susana Ribot

Copy (supplied by) Bet Barceló

Editing Mar Valls, Alberto Hernández

Graphic editing Alberto Hernández

Layout Clara Miralles, João Neves

Copy editor Stuart Franklin

Infographics Sol90Images

Photography AGE Fotostock, Getty Images, Cordon Press/Corbis, ACI, Album, Cordon Press/Granger

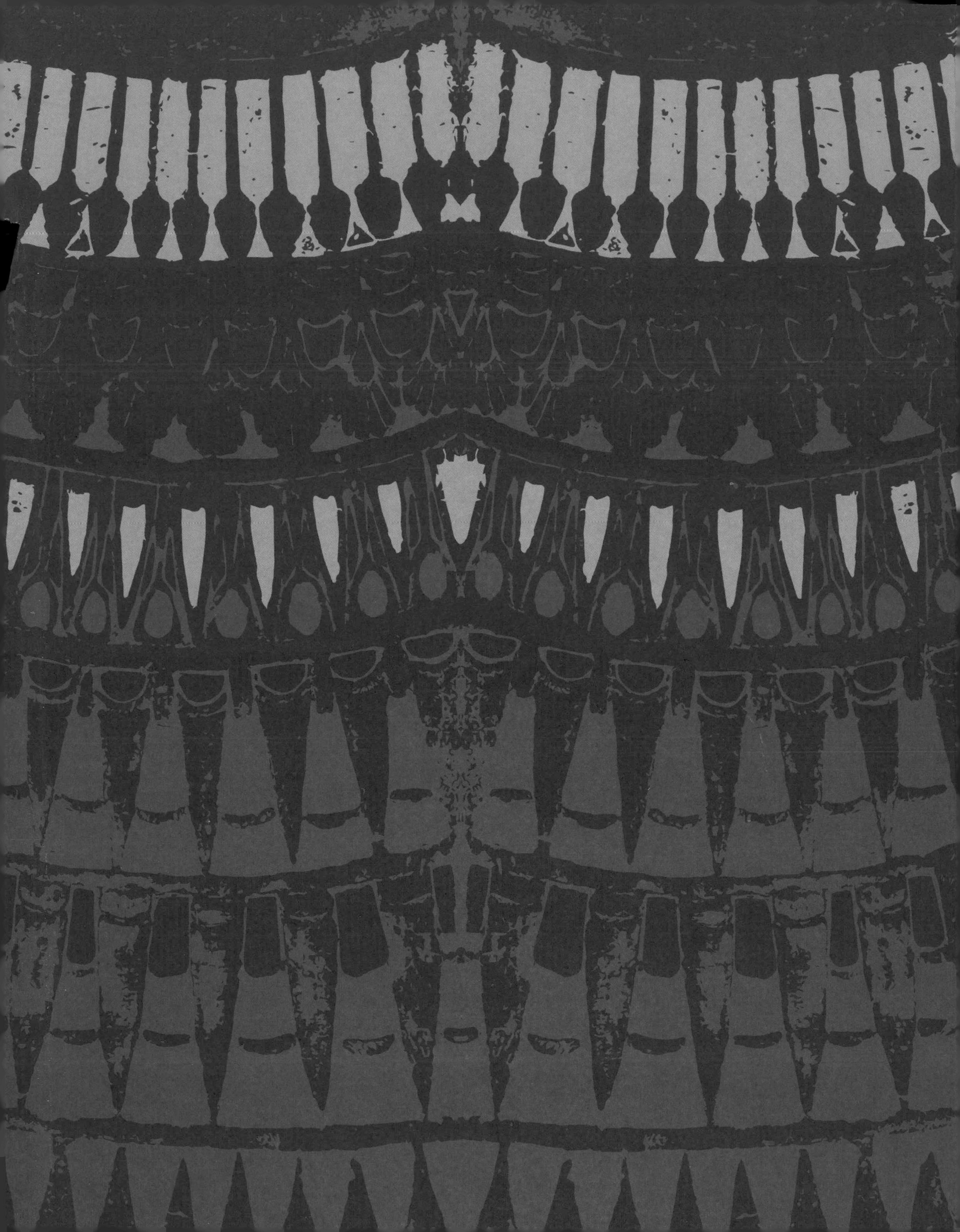